CAMBRIDGE MONOGRAPHS ON
APPLIED AND COMPUTATIONAL
MATHEMATICS

Series Editors
M. ABLOWITZ, S. DAVIS, J. HINCH,
A. ISERLES, J. OCKENDON, P. OLVER

26   A Practical Guide to the Invariant Calculus

The *Cambridge Monographs on Applied and Computational Mathematics* series reflects the crucial role of mathematical and computational techniques in contemporary science. The series publishes expositions on all aspects of applicable and numerical mathematics, with an emphasis on new developments in this fast-moving area of research.

State-of-the-art methods and algorithms as well as modern mathematical descriptions of physical and mechanical ideas are presented in a manner suited to graduate research students and professionals alike. Sound pedagogical presentation is a prerequisite. It is intended that books in the series will serve to inform a new generation of researchers.

Also in this series:

A Practical Guide to the Invariant Calculus

ELIZABETH LOUISE MANSFIELD
University of Kent, Canterbury

CAMBRIDGE
UNIVERSITY PRESS

University Printing House, Cambridge CB2 8BS, United Kingdom

One Liberty Plaza, 20th Floor, New York, NY 10006, USA

477 Williamstown Road, Port Melbourne, VIC 3207, Australia

314-321, 3rd Floor, Plot 3, Splendor Forum, Jasola District Centre, New Delhi - 110025, India

103 Penang Road, #05-06/07, Visioncrest Commercial, Singapore 238467

Cambridge University Press is part of the University of Cambridge.

It furthers the University's mission by disseminating knowledge in the pursuit of education, learning and research at the highest international levels of excellence.

www.cambridge.org
Information on this title: www.cambridge.org/9780521857017

First published 2010

A catalogue record for this publication is available from the British Library

ISBN 978-0-521-85701-7 Hardback

Contents

Preface

I first became enamoured of the Fels and Olver formulation of the moving frames theory when it helped me solve a problem I had been thinking about for several years. I set about reading their two 50-page papers, and made a 20-page handwritten glossary of definitions. I was lucky in that I was able to ask Peter Olver many questions and am eternally grateful for the answers.

I set about solving the problems that interested me, and realised there were so many of them that I could write a book. I also wanted to share my amazement at just how powerful the methods were, and at the essential simplicity of the central idea. What I have tried to achieve in this book is a discussion rich in examples, exercises and explanations that is largely accessible to a graduate student, although access to a professional mathematician will be required for some parts. I was extremely fortunate to have six students read through various drafts from the very beginning. The comments and hints they needed have been incorporated, and I have not hesitated to put in a discussion, example, exercise or hint that might be superfluous to a professional.

There is a fair amount of original material in this book. Even though some of the problems addressed here have been solved using moving frames already, I have re-proved some results to keep both solution methods and proofs within the domain of the mathematics developed here. I love coming up with simpler solutions. In particular, the variational methods developed in Chapter 7 are my own. The theorem on moving frames and Noether's Theorem, which was discovered and proved with Tania Gonçalves, particularly pleases me. The application of moving frames to the solution of invariant ordinary differential equations is also new. I was particularly chuffed to solve the Chazy equation using relatively simple calculations, see Chapter 6. Theorem 5.2.4 allowing one to write down the curvature matrices in terms of a matrix representation of the frame was published earlier in Mansfield and van der Kamp (2006), and there are some fun exercises giving new applications. Finally some minor (and not so minor) errors in the original papers have been corrected.

The natural setting of the problems that interested me did not fit well with the language of differential geometry in which all discussions of moving frames were couched, so I set about casting the calculations into ordinary undergraduate calculus in order to explain it in my papers and then to teach it to my students. It was clear that a major benefit of Fels and Olver's formulation of the central concept was that it actually freed the moving frame method from the confines of differential geometry; that it could apply equally well to differential difference problems, to discrete problems, to all kinds of numerical approximations and so on. In any event, there are serious problems with that language as an expository tool.[†] Thus when I decided to write up my notes into a book, I was clear in my own mind that I was not going to use the exterior calculus as the primary expository language. Nevertheless, it is important to have available coordinate-free expressions if we are not to suffer 'death by indices'. What I wanted was a language that offered concrete models of objects like smooth functions, vectors and vector fields, capable of use in both finite and infinite dimensional spaces, that was linked in an open, explicit and well-defined way to multivariable calculus, and for which there was a good literature where the central significant theorems were proved properly. The language I needed, and use, is that of Differential Topology. I learned this subject twice, first at the University of Sydney in lectures given by M. J. Field, and then at the University of Wisconsin, Madison, in a year long course given by Dennis Stowe. I am extremely grateful to them both. The notation and language that I use in this book is what they both independently taught me, which has stood me in good stead my whole career.

A huge contribution to the theory of moving frames, as they can be studied rigorously in a symbolic computation environment, has been made by Evelyne Hubert. One of the main benefits of the Fels and Olver formulation of moving frames is that much of the calculation can be done symbolically in a computer algebra environment. The fact that one can have a symbolic calculus of invariants, without actually solving for the frame, is what turns this theory from the merely beautiful to the both beautiful and useful; this is the hallmark of the best mathematics. From the point of view of *rigorous* symbolic computation, though, there were problems, in particular with the need to invoke the implicit function theorem because this is a non-constructive step. Evelyne Hubert and Irina Kogan (Hubert and Kogan, 2007a) provide algebraic foundations to the moving frame method for the construction of local invariants and present a

---

[†] Don't get me started.

parallel algebraic construction that produces algebraic invariants together with the relations they satisfy. They then show that the algebraic setting offers a computational solution to the original differential geometric construction.

A second problem solved by Evelyne Hubert was the lack of a theory to analyse the differential systems resulting from invariantisation, since these involve non-commuting differential operators. Indeed, none of the edifice of mathematics that had been produced to study over determined differential systems rigorously was applicable, although an equivalent theory was needed for the applications (Mansfield, 2001). In a beautiful exposition (Hubert, 2005), the web of difficulties was pulled apart, the necessary concepts and results were lined up in order, and the required theory was developed.

A third problem solved by Hubert was that of proving that a certain small, finite set of syzygies, or differential relations satisfied by the invariants, generated the complete set of syzygies (Hubert, 2009a). This was important since the theorem written down by Fels and Olver turned out to be false in general.

Finally, Hubert finds a set of generators of the algebra of differential invariants that are not only simple to calculate but simple to conceptualise (Hubert, 2009b).

To give an exposition of these papers at the level I wrote this volume would require another volume, with a substantial expository section on over determined systems. However, the papers are accessible and I commend them to the reader.

When I started to view the material from the point of view of my target audience, primarily people wanting to use the methods but not having learnt (nor wanting to learn) Differential Geometry, and also graduate students, I came to realise that the subject involves a significant range of mathematics that could not realistically be assumed knowledge. Brief but necessary remarks on topics from transversality to foliations to jet bundles, and on calculations in Lie algebras and the variational calculus, all swelled to much longer expository sections than I anticipated. One central classical theorem for which I could not find a decent modern exposition of the proof was Frobenius' Theorem, so I have outlined the proof in a series of exercises. The outline is based on that given in lectures at the University of Wisconsin, Madison, by Dennis Stowe, to whom I acknowledge my debt.

In writing this book I have tried to steer a course through the material that is both honest and pragmatic. If being rigorous would have involved too long a detour, I chose computation of examples and discussion over rigour; it is more insightful to discuss the meaning rather than the proof of a result when there is a good text that can be consulted for further reading. Where I do give a proof, though, I aimed for the proof to follow rigorously from the established base of knowledge. Interestingly, sometimes not even the cleanest, simplest proofs reveal the inner truth: the full understanding of theorems can only be achieved after a range of examples can be computed. I give many exercises, hints, and details in my own calculations to help my readers to two levels of computational expertise: first, to be able to correctly work simple examples that can be done by hand or performed interactively with a computer algebra package, and second, to be able to write a computer program to do his or her own larger examples.

I wish to thank Peter Olver, Evelyne Hubert, Peter Hydon and Francis Valiquette, who sent me comments. I had some great discussions with Gloria Marí Beffa, resulting in several beautiful examples that are described in the text. Peter van der Kamp's insistence on in-depth detail for his own understanding of moving frames made this a much better book. Tania Gonçalves, Richard Hoddinott, Jun Zhao and Andrew Wheeler worked through the exercises; readers can thank them for the hints and for amplified discussions in various places. I road tested the very first set of notes on Emma Berry and Andrew Martin whose comments helped me see things from my target audience's point of view.

As ever, I wish to thank my dear husband Peter Clarkson who supported me in a million different ways when the going got tough. I have faced and overcome some extraordinary obstacles in order to have a mathematical career; I have my father Dr Colin Mansfield, my PhD thesis supervisor Dr Edward Fackerell (Sydney), and my mentor Professor Arieh Iserles (Cambridge) to thank for their extraordinary timely support. Words cannot express how lucky and how grateful I feel to have such stalwart friends and fellow travellers.

The author would like to acknowledge the Engineering and Physical Sciences Research Council (UK) grant, 'Symmetric variational problems' EP/E001823/1.

Introduction to invariant and equivariant problems

The curve completion problem

Consider the 'curve completion problem', which is a subproblem of the much more complex 'inpainting problem'. Suppose we are given a partially obscured curve in the plane, as in Figure 0.1, and we wish to fill in the parts of the curve that are missing. If the missing bit is small, then a straight line edge can be a cost effective solution, but this does not always give an aesthetically convincing look. Considering possible solutions to the curve completion problem (Figure 0.2), we arrive at three requirements on the resulting curve:

- it should be sufficiently smooth to fool the human eye,
- if we rotate and translate the obscured curve and then fill it in, the result should be the same as filling it in and then rotating and translating,
- it should be the 'simplest possible' in some sense.

The first requirement means that we have boundary conditions to satisfy as well as a function space in which we are working. The second means the formulation of the problem needs to be 'equivariant' with respect to the standard action of the Euclidean group in the plane, as in Figure 0.3. This condition arises naturally: for example, if the image being repaired is a dirty photocopy, the result should not depend on the angle at which the original is fed into the photocopier.

All three conditions can be satisfied if we require the resulting curve to be such as to minimise an integral which is invariant under the group action,

$$\int L(s, \kappa, \kappa_s, \dots) \, \mathrm{d}s, \qquad (0.1)$$

1

Figure 0.1 A curve in the plane with occlusions.

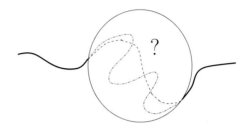

Figure 0.2 Which infilling is best?

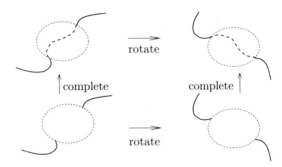

Figure 0.3 The solution is equivariant.

where s is arc length and κ the Euclidean curvature,

$$\kappa = \frac{u_{xx}}{(1+u_x^2)^{3/2}}, \qquad \frac{\mathrm{d}}{\mathrm{d}s} = \frac{1}{\sqrt{1+u_x^2}}\frac{\mathrm{d}}{\mathrm{d}x} \qquad (0.2)$$

and $\mathrm{d}s = \sqrt{1+u_x^2}\,\mathrm{d}x$.

The theory of the Calculus of Variations is about finding curves that minimise integrals such as equation (0.1), and the most famous Lagrangian in this family is

$$\mathcal{L}[u] = \int \kappa^2 \,\mathrm{d}s. \qquad (0.3)$$

The main theorem in the Calculus of Variations is that the minimising curves satisfy a differential equation called the Euler–Lagrange equation. There are quite a few papers and even textbooks that either 'prove' or assume the wrong Euler–Lagrange equation for (0.3), namely that the minimising curve is a circle, that is, satisfying $\kappa = c$. The correct result, calculated by Euler himself, is that the curvature of the minimising curve satisfies

$$\kappa_{ss} + \frac{1}{2}\kappa^3 = 0, \tag{0.4}$$

which is solved by an elliptic function. Solutions are called 'Euler's elastica' and have many applications. See Chan *et al.* (2002) for a discussion relevant to the inpainting problem.

While Euler–Lagrange equations can be found routinely by symbolic computation packages, and then rewritten in terms of historically known invariants, this process reveals little to nothing of why the Euler–Lagrange equation has the terms and features it does. The motivating force behind Chapter 7 was to bring out and understand the structure of Euler–Lagrange equations for variational problems where the integrand, called a Lagrangian, is invariant under a group; the groups relevant here are not finite groups, but *Lie groups*, those that can be parametrised by real or complex numbers, such as translations and rotations.

One of the most profound theorems of the Calculus of Variations is Noether's Theorem, giving formulae for first integrals of Euler–Lagrange equations for Lie group invariant Lagrangians. Most Lagrangians arising in physics have such an invariance; the laws of nature typically remain the same under translations and rotations, also pseudorotations in relativistic calculations, and so on, and thus Noether's Theorem is well known and much used.

If one calculates Noether's first integrals for the variational problem (0.3), the result can be written in the form,

$$\begin{pmatrix} c_1 \\ c_2 \\ c_3 \end{pmatrix} = \begin{pmatrix} \dfrac{1}{\sqrt{1+u_x^2}} & -\dfrac{u_x}{\sqrt{1+u_x^2}} & 0 \\[2mm] \dfrac{u_x}{\sqrt{1+u_x^2}} & \dfrac{1}{\sqrt{1+u_x^2}} & 0 \\[2mm] \dfrac{xu_x - u}{\sqrt{1+u_x^2}} & \dfrac{uu_x + x}{\sqrt{1+u_x^2}} & 1 \end{pmatrix} \begin{pmatrix} -\kappa^2 \\ -2\kappa_s \\ 2\kappa \end{pmatrix} \tag{0.5}$$

where the c_i are the constants of integration. The first component comes from translation in x, the second from translation in u and the third from rotation in the (x, u) plane about the origin. The 3×3 matrix appearing in (0.5), which I denote here by $B(x, u, u_x)$, has a remarkable property. If one calculates the induced action of the group of rotations and translations in the plane, that is, the special Euclidean group $SE(2)$, on B, componentwise, then one has

$$B(g \cdot x, g \cdot u, g \cdot u_x) = R(g)B(x, u, u_x), \quad \text{for all } g \in SE(2)$$

where $R(g)$ is a particular matrix representation of $SE(2)$ called the Adjoint representation. In other words, $B(x, u, u_x)$ is *equivariant* with respect to the group action, and is thus an equivariant map from the space with coordinates $(x, u, u_x, u_{xx}, \dots)$ to $SE(2)$. The equivariance can be used to understand how the group action takes solutions of the Euler–Lagrange equations to solutions.

Equivariant maps are, in fact, the secret to success for the invariant calculus. They are denoted as a 'moving frame' and are the central theme of Chapter 4. In Chapter 7 we prove results that give the structure of both the Euler–Lagrange equations and the set of first integrals for invariant Lagrangians, using the symbolic invariant calculus developed in Chapters 4 and 5. The fact that the formula for Noether's Theorem yields the very map required to establish the symbolic invariant calculus, used in turn to understand the structure of the results, continues to amaze me.

Curvature flows and the Korteweg–de Vries equation

Consider the group of 2×2 real matrices with determinant 1, called $SL(2)$, which we write as

$$SL(2) = \left\{ \begin{pmatrix} a & b \\ c & d \end{pmatrix} \mid ad - bc = 1 \right\}.$$

We are interested in actions of this group on, say, curves in the (x, u) plane, that evolve in time, so our curves are parametrised as $(x, t, u(x, t))$. Suppose for $g \in SL(2)$ we impose that the group acts on curves via the map

$$g \cdot x = x, \qquad g \cdot t = t, \qquad g \cdot u = \frac{au + b}{cu + d}.$$

Using the chain rule, we can induce an action on u_x and higher derivatives, as

$$g \cdot u_x = \frac{\partial(g \cdot u)}{\partial(g \cdot x)} = \frac{u_x}{(cu + d)^2},$$

and

$$g \cdot u_{xx} = \frac{\partial(g \cdot u_x)}{\partial(g \cdot x)}$$

and so on. It is then a well-established historical fact that the lowest order invariants are

$$W = \frac{u_t}{u_x}, \qquad V = \frac{u_{xxx}}{u_x} - \frac{3}{2}\frac{u_{xx}^2}{u_x^2} := \{u; x\}.$$

The invariant V is called the Schwarzian derivative of u and is often denoted as $\{u; x\}$. This derivative featured strongly in the differential geometry of a bygone era; it is used today in the study of integrable systems. The reason is as follows. The invariants V and W are functionally independent, but there is a differential identity or syzygy,

$$\frac{\partial}{\partial t}V = \underbrace{\left(\frac{\partial^3}{\partial x^3} + 2V\frac{\partial}{\partial x} + V_x \right)}_{\mathcal{H}} W.$$

The operator $\mathcal{H}$ appearing in this equation is one of the two Hamiltonian operators for the Korteweg–de Vries equation, see Olver (1993), Example 7.6, with $V = u/3$. Thus, if $W = V$, that is if $u_t = u_x\{u; x\}$, then $V(x, t)$ satisfies the Korteweg–de Vries equation.

In fact there are many examples like this, where syzygies between invariants give rise to pairs of partial differential equations that are integrable, with one of the pair being in terms of the invariants of a given smooth group action. Another example of such a pair is the vortex filament equation and the non-linear Schrödinger equation. In that case, the group action is the standard action of the group of rotations and translations in $\mathbb{R}^3$. We refer to Mansfield and van der Kamp (2006) and to Marí Beffa (2004, 2007, 2008a, 2008b) for more information.

The essential simplicity of the main idea

For many applications, what seems to be wanted is the following:

> given the smooth group action, derive the invariants and their syzygies *algorithmically*, that is, without prior knowledge of 100 years of differential geometry, and with *minimal effort*.

To show the essential simplicity of the main idea, we consider a simple set of transformations of curves $(x, u(x))$ in the plane given by

$$x \mapsto \tilde{x} = \lambda x + k, \qquad u \mapsto \tilde{u} = \lambda u, \qquad \lambda \neq 0. \tag{0.6}$$

The induced action on tangent lines to the curves is given by the chain rule:

$$u_x \mapsto \frac{d\widetilde{u}}{d\widetilde{x}} = \frac{d\widetilde{u}}{dx}\left(\frac{d\widetilde{x}}{dx}\right)^{-1} = \frac{\lambda u_x}{\lambda} = u_x$$

and so u_x is an invariant. Continuing, we obtain

$$u_{xx} \mapsto \frac{u_{xx}}{\lambda}, \qquad u_{xxx} \mapsto \frac{u_{xxx}}{\lambda^2}$$

and so on. Of course, in this simple example, we can see what the invariants have to be. But let us pretend we do not for some reason, and derive a set of invariants.

The basic idea is to solve two equations for the two parameters λ and k. If we take $\widetilde{x} = 0$ and $\widetilde{u} = 1$, we obtain

$$\lambda = \frac{1}{u}, \qquad k = -\frac{x}{u}. \tag{0.7}$$

We give these particular values of the parameters the grand title 'the frame'. If we now evaluate the images of $u_{xx}, u_{xxx}, \ldots$ under the mapping, with λ and k given by the frame parameters in equation (0.7), we obtain

$$u_{xx} \mapsto \frac{u_{xx}}{\lambda} \mapsto u u_{xx}, \qquad u_{xxx} \mapsto \frac{u_{xxx}}{\lambda^2} \mapsto u^2 u_{xxx}, \qquad \ldots.$$

We now observe that the final images of our maps are all invariants. Indeed,

$$u^2 u_{xxx} \mapsto (\lambda u)^2 \left(\frac{u_{xxx}}{\lambda^2}\right) = u^2 u_{xxx}$$

and so on. The method of 'solve for the frame, then back-substitute' has produced an invariant of every order, specifically

$$I_n = u^{n-1} \underbrace{u_{x x \ldots x}}_{n \text{ terms}}.$$

It is easy to show that any invariant can be expressed in terms of the I_n. Indeed, if $F(x, u, u_x, \ldots)$ is an invariant, then

$$F(x, u, u_x, u_{xx}, \ldots) = F\left(\lambda x + k, \lambda u, u_x, \frac{u_{xx}}{\lambda}, \ldots\right) \tag{0.8}$$

for all λ and k. If I use the 'frame' values of the parameters in equation (0.8), I obtain

$$F(x, u, u_x, u_{xx}, \ldots) = F(0, 1, u_x, u u_{xx}, \ldots) = F(0, 1, I_1, I_2, \ldots).$$

Since any invariant at all can be written in terms of the I_n, we have what is called a *generating set* of invariants.

But that is not all. If I use the same approach on the derivative operator

$$\frac{\mathrm{d}}{\mathrm{d}x} \mapsto \frac{\mathrm{d}}{\mathrm{d}\widetilde{x}} = \left(\frac{\mathrm{d}\widetilde{x}}{\mathrm{d}x}\right)^{-1}\frac{\mathrm{d}}{\mathrm{d}x} = \lambda\frac{\mathrm{d}}{\mathrm{d}x} \mapsto u\frac{\mathrm{d}}{\mathrm{d}x}$$

then the final result,

$$\mathcal{D} = u\frac{\mathrm{d}}{\mathrm{d}x}$$

is invariant, that is,

$$\mathcal{D} \mapsto \widetilde{u}\frac{\mathrm{d}}{\mathrm{d}\widetilde{x}} = u\frac{\mathrm{d}}{\mathrm{d}x} = \mathcal{D}.$$

Differentiating an invariant with respect to an invariant differential operator must yield an invariant, and indeed we obtain

$$\mathcal{D}I_1 = I_2, \qquad \mathcal{D}I_2 = I_3 + I_1 I_2 \tag{0.9}$$

and so on.

Equations of the form (0.9) are called *symbolic differentiation formulae*. The major advance made by Fels and Olver (Fels and Olver, 1998, 1999) was to find a way to obtain equations (0.9) without knowing the frame, but only the equations used to define the frame, which in this case were $\widetilde{x} = 0, \widetilde{u} = 1$.

If we now look at a matrix form of our mapping,

$$\begin{pmatrix} \widetilde{x} \\ \widetilde{u} \\ 1 \end{pmatrix} = \begin{pmatrix} \lambda & 0 & k \\ 0 & \lambda & 0 \\ 0 & 0 & 1 \end{pmatrix}\begin{pmatrix} x \\ u \\ 1 \end{pmatrix}$$

and evaluate the matrix of parameters on the frame, we obtain 'the matrix form of the frame',

$$\varrho(x, u) = \begin{pmatrix} \dfrac{1}{u} & 0 & -\dfrac{x}{u} \\ 0 & \dfrac{1}{u} & 0 \\ 0 & 0 & 1 \end{pmatrix}.$$

Going one step further, if we act on this matrix $\varrho(x, u)$, we obtain

$$
\varrho(\widetilde{x}, \widetilde{u}) =
\begin{pmatrix}
\dfrac{1}{\widetilde{u}} & 0 & -\dfrac{\widetilde{x}}{\widetilde{u}} \\[2mm]
0 & \dfrac{1}{\widetilde{u}} & 0 \\[2mm]
0 & 0 & 1
\end{pmatrix}
$$

$$
=
\begin{pmatrix}
\dfrac{1}{u} & 0 & -\dfrac{x}{u} \\[2mm]
0 & \dfrac{1}{u} & 0 \\[2mm]
0 & 0 & 1
\end{pmatrix}
\begin{pmatrix}
\dfrac{1}{\lambda} & 0 & -\dfrac{k}{\lambda} \\[2mm]
0 & \dfrac{1}{\lambda} & 0 \\[2mm]
0 & 0 & 1
\end{pmatrix}
$$

$$
= \varrho(x, u)
\begin{pmatrix}
\lambda & 0 & k \\
0 & \lambda & 0 \\
0 & 0 & 1
\end{pmatrix}^{-1} .
$$

What this result means is that 'the frame' is *equivariant* with respect to the mapping (0.6).

The miracle is that the entire symbolic calculus can be built from the equivariance of the frame and ordinary multivariable calculus, even if you do not know the frame explicitly, that is, even if you cannot solve the equations giving the frame for the parameters.

The one caveat is that not any old mapping involving parameters can be studied this way; the mapping (0.6) is in fact a *Lie group action*, where the Lie group is the set of 3×3 matrices

$$
\left\{
\begin{pmatrix}
\lambda & 0 & k \\
0 & \lambda & 0 \\
0 & 0 & 1
\end{pmatrix} \mid \lambda, k \in \mathbb{R}, \lambda > 0
\right\}
$$

(amongst other representations) which is closed under multiplication and inversion.

Not all group actions are linear like (0.6), and since we do not need to assume linearity for any of the theory to be valid, we do not assume it. However, often the version of a theorem assuming a linear action is easier to state and prove, and so we tend to do both.

Overview of this book

My primary aim in writing this book was to bring the theory and applications of moving frames to an audience not wishing to learn Differential Geometry first, to show how the calculations can be done using primarily undergraduate calculus, and to provide a discussion of a range of applications in a fully detailed way so that readers can do their own calculations without undue headscratching.

The main subject matter is, first and foremost, smooth group actions on smooth spaces. Surprisingly, this includes applications to many seemingly discrete problems. The groups referred to in this book are *Lie groups*, groups that depend on real or complex parameters. In Chapter 1 we discuss the basic notions concerning Lie groups and their actions, particularly their actions as prolonged to derivative terms. Since there is a wealth of excellent texts on this topic, we cruise through the examples, calculations and basic definitions, introducing the main examples I use throughout.

The following two chapters give foundational material for Lie theory as I use and need it for this book. I could not find a good text with exactly what was needed, together with suitable examples and exercises, so I have written this myself, proving everything from scratch. While I imagine most readers will only refer to them as necessary, hopefully others will be inspired to learn more Differential Topology and Lie Theory from texts dedicated to those topics.

In Chapter 2, I discuss how multivariable calculus extends to a calculus on Lie groups; this is mostly an introduction to standard Differential Topology for the particular cases of interest, and a discussion of the central role played by one parameter subgroups. The point of view taken in Differential Topology, on 'what is a vector' and 'what is a vector field', is radically different to that taken in Differential Geometry. The first theory bases the notion of a vector on a path, the second on the algebraic notion of a derivation acting on functions. There are serious problems with a *definition* of a vector field as a derivation.[†] On the other hand, the notion of a vector as a path in the space, which can be differentiated at its distinguished point in coordinates, is a powerful, all purpose, take anywhere idea that has a clear and explicit link to standard multivariable calculus. Further, anyone who has witnessed a leaf being carried by water, or a speck of dust being carried by the wind, has already developed the necessary corresponding intuitive notion. Armed with the clear and useful notion of a vector as a path, everything we need can be proved from the theorem guaranteeing the existence and uniqueness of a solution to first order differential systems. So as to give

---

[†] Not the least problem is that the chain rule needed for the transformation of vectors does not follow from this definition alone, which can apply equally well to strictly algebraic objects.

those well versed in one language insight into the other, we give some links between the two sets of ideas and the relevant notations.

In the second chapter on the foundations of Lie theory, Chapter 3, we discuss the Lie bracket of vector fields and Frobenius' Theorem and from there, the Lie algebra, the Lie bracket, and the Adjoint and adjoint actions. The two quite different appearances of the formulae for the Lie bracket, for matrix groups and transformation groups, are shown to be instances of the one general construction, which in turn relies on the Lie bracket of vector fields in $\mathbb{R}^n$. While many authors simply give the two different formulae as definitions, I was not willing to do that for reasons I make clear in the introduction to that chapter.

Chapters 4 and 5 are the central chapters of the book. The key idea underlying the symbolic invariant calculus is a formulation of a moving frame as an equivariant map from space M on which the group G acts, to the group itself. When one can solve for the frame, one has explicit invariants and invariant differential operators. When one cannot solve for the frame, then one has symbolic invariants and invariant differential operators. This is the topic of Chapter 4, which introduces the distinguished set of symbolic invariants and symbolic invariant differentiation operators used throughout the rest of the book. Chapter 5 continues the main theoretical development to discuss the differential relations or syzygies satisfied by the invariants, and introduces the curvature matrices. These are well known in differential geometry, and we discuss the famous Serret–Frenet frame, but they have other applications; in particular, they can be used to solve numerically for the frame. Both chapters have sections detailing various applications and further developments; sections designated by a star, *, can be omitted on a first reading.

From this firm theoretical foundation, a host of applications can be described. The two most developed applications in this book are to solving invariant ordinary differential equations, and to the Calculus of Variations. In fact, there is a long history of using smooth group actions to solve invariant ordinary differential equations; normally one would think of this theory as a success story, with little more to say. However, we describe in Chapter 6 just how much more can be achieved with the new ideas. Similarly, the Calculus of Variations is a classical subject that one might think of as fully mature. In Chapter 7, the use of the new ideas throws substantial light on the *structure* of the known results when invariance under a smooth group action is given.

The three applications that pleased me the most were solving the Chazy equation, finding the equations for a free rigid body without any mysterious concepts, and the final theorem of the book, showing the structure of the first integrals given by Noether's Theorem. All three came out of trying to develop interesting exercises for this book.

Other applications to discrete problems, to functional approximations or numerical integration problems, remain to be developed in the research literature to the same level as those I have written about here. These are some of my favourite applications, but I ran out of time and space. Another application I wanted to include was the extension of the Fels and Olver reformulation of the moving frame to pseudogroups (Cheh *et al.*, 2005; Olver and Pohjanpelto, 2008; Shemyakova and Mansfield, 2008). This, too, will have to wait for a second volume.

How to read this book . . .

. . . is with pencil, paper and symbolic computation software. The only way to see the magic is to do it.

1
Actions galore

1.1 Introductory examples

A *symmetry* of a function is a coordinate transformation which leaves the function invariant. Consider the function $r(x, y, z) = \sqrt{x^2 + y^2 + z^2}$ which is the distance of a point $(x, y, z) \in \mathbb{R}^3$ from the origin. This function is invariant under rotations about the origin. Indeed, consider a rotation matrix $R : \mathbb{R}^3 \to \mathbb{R}^3$. First note that $R^T R = I_3$, the 3×3 identity matrix. Setting $\mathbf{x} = (x, y, z)^T$ and $\mathbf{X} = (X, Y, Z)^T = R\mathbf{x}$, then

$$
\begin{aligned}
x^2 + y^2 + z^2 &= \mathbf{x}^T \mathbf{x} \\
&= \mathbf{x}^T R^T R \mathbf{x} \\
&= (R\mathbf{x})^T R\mathbf{x} \\
&= \mathbf{X}^T \mathbf{X} \\
&= X^2 + Y^2 + Z^2.
\end{aligned}
$$

Next consider $\tau = \tau(x, y, z, t)$ known as 'proper time', arising in the theory of special relativity, given by

$$
c^2 \tau^2 = c^2 t^2 - x^2 - y^2 - z^2
$$

where c is the speed of light, assumed to be a constant. Symmetries of τ include rotations in the space variables (x, y, z). Since both $x^2 + y^2 + z^2$ and t are invariant under rotations in the space variables, τ is invariant.

Pseudorotations are also symmetries of τ. These can be described in a variety of ways. If we set y and z to be invariant and restrict our attention to (x, t) space, then a pseudorotation $g(v)$ is given by

$$
\begin{pmatrix} x \\ t \end{pmatrix} \mapsto g(v) \cdot \begin{pmatrix} x \\ t \end{pmatrix},
$$

where $g(v)$ is the matrix,

$$g(v) = \begin{pmatrix} \dfrac{1}{\sqrt{1 - v^2/c^2}} & -\dfrac{v}{\sqrt{1 - v^2/c^2}} \\ -\dfrac{v/c^2}{\sqrt{1 - v^2/c^2}} & \dfrac{1}{\sqrt{1 - v^2/c^2}} \end{pmatrix}. \tag{1.1}$$

The parameter v is the velocity of the new coordinate system relative to the first, and is assumed to be in the range $v \in (-c, c)$. Note the pseudorotation varies smoothly with v. We have that the inverse transformation $(g(v))^{-1}$ equals $g(-v)$ and the identity transformation is $g(0)$, that is, with $v = 0$. The composition of velocities is non-linear, however; indeed,

$$g(v) \cdot g(w) = g(\xi(v, w)) \tag{1.2}$$

where

$$\xi(v, w) = \frac{v + w}{1 + vw/c^2}. \tag{1.3}$$

If $v, w \in (-c, c)$ then $\xi(v, w) \in (-c, c)$, so the group composition law respects the restriction on the parameters. Another representation of the set $\{g(v) \mid v \in (-c, c)\}$ is $\{\bar{g}(\alpha) \mid \alpha \in \mathbb{R}\}$, where

$$\bar{g}(\alpha) = \begin{pmatrix} \cosh \alpha & c \sinh \alpha \\ \frac{1}{c} \sinh \alpha & \cosh \alpha \end{pmatrix}.$$

The relationship between the two representations is

$$g(v) \mapsto \bar{g}(\alpha), \qquad \tanh(\alpha) = -v/c. \tag{1.4}$$

The composition law is simpler in the second representation;

$$\bar{g}(\alpha) \cdot \bar{g}(\beta) = \bar{g}(\alpha + \beta). \tag{1.5}$$

Exercise 1.1.1 Show equation (1.2) holds with $\xi(v, w)$ given in (1.3), and that equation (1.5) also holds. Show that the relationship (1.4) sends one composition law to the other. Hint: set $\cosh \alpha = 1/\sqrt{1 - v^2/c^2}$, $\sinh \alpha = -v/(c\sqrt{1 - v^2/c^2})$ and use the addition rule for tanh.

Similar remarks hold for pseudorotations in the (y, t) and (z, t) planes.

The set of all linear coordinate transformations of (x, y, z, t) space that can be written as a finite composition of rotations and pseudorotations is a subset of the Lorentz group.

Definition 1.1.2 Let Λ be the diagonal matrix with diagonal entries being $(-1/c^2, 1, 1, 1)$. The *Lorentz group* is defined to be the set of 4×4 real invertible matrices A such that

$$A^T \Lambda A = \Lambda.$$

Definition 1.1.3 A *group* G is a set such that

- there is a map, called the product map,

$$\mu : G \times G \to G.$$

  We refer to the fact that the product μ maps *into G* as the *closure property*.
- The map μ satisfies the *associative law*,

$$\mu(g_1, \mu(g_2, g_3)) = \mu(\mu(g_1, g_2), g_3).$$

- There is an element e called the *identity* such that

$$\mu(e, g) = \mu(g, e) = g, \quad \text{for all } g \in G$$

  and,
- each $g \in G$ has an *inverse*, denoted g^{-1}, such that

$$\mu(g^{-1}, g) = \mu(g, g^{-1}) = e.$$

We write the group as (G, μ) when we need to specify the product. Where the product is clear from the context, we write $\mu(g_1, g_2)$ as $g_1 g_2$.

Definition 1.1.4 If the order of multiplication does not matter, so that $g_1 g_2 = g_2 g_1$ for all $g_1, g_2 \in G$, we say that G is a *commutative* group.

Standing assumption   The groups we consider in this book are **Lie groups**, groups whose elements depend in a smooth way on parameters, which in this book are real or complex numbers. Parameters from other fields such as the quarternions can also be used. A generic group element of a Lie group is written as $g = g(a_1, \ldots, a_r)$ when we wish to highlight the independent parameters on which g depends. A rigorous definition of a Lie group is given in the next chapter.

Lie groups can arise as symmetry groups of sets of functions. If the set of functions is the solution set of a differential system, then we speak of the Lie group of symmetries of the system. More generally, we consider symmetries of differentiable geometric structures. In this chapter we define the basic terms we use and give a range of examples that will be used in later chapters.

The simplest Lie group is the set of real numbers with addition as the group 'product', $(\mathbb{R}, +)$, so $\mu(x, y) = x + y$. The identity element is zero, and the inverse of an element is its negative. A second Lie group is $(\mathbb{R}^+, \cdot)$, where $\mathbb{R}^+$ denotes positive numbers and the group product is the standard multiplication. These two groups are essentially the same group because there is an invertible map

$$f : (\mathbb{R}, +) \to (\mathbb{R}^+, \cdot)$$

that maps 'products' in $(\mathbb{R}, +)$ to the corresponding product in $(\mathbb{R}^+, \cdot)$,

$$f(x + y) = f(x)f(y).$$

Of course, f is the exponential map. We formalise this notion of 'being the same group' by defining an isomorphism.

Definition 1.1.5 Two groups (G, μ_G) and (H, μ_H) are said to be *isomorphic*, if there is an invertible map

$$f : G \to H$$

satisfying

$$f(\mu_G(g_1, g_2)) = \mu_H(f(g_1), f(g_2)), \qquad \text{for all } g_1, g_2 \in G.$$

We write $G \approx H$ and say that 'f respects the group product'.

The next simplest Lie groups to describe are *matrix groups*. These are sets of square matrices and the product is always the usual matrix multiplication. Since this product is known to be associative, to test whether a set of matrices is a group, we need to check

(i) *closure*, that is, the product of two elements in the set is itself in the set,
(ii) that the identity matrix of the relevant dimension is in the set, and
(iii) not only that elements of the set are invertible, that is, their determinants are non-zero, but also that their inverses are in the set.

We denote the set of all $n \times n$ matrices whose components are real numbers by $M_n(\mathbb{R})$, and the set of invertible $n \times n$ matrices whose entries are real numbers by $GL(n, \mathbb{R})$. Similarly, the set of invertible $n \times n$ matrices whose entries are elements of the field $\mathbb{F}$ is denoted $GL(n, \mathbb{F})$ or simply $GL(n)$ if the field is obvious from the context. All matrix groups are subgroups of $GL(n, \mathbb{F})$ for some $\mathbb{F}$ and some n.

Remark 1.1.6 Another notation that we will use for $M_n(\mathbb{F})$ is $\mathfrak{gl}(n, \mathbb{F})$, for reasons that will be become apparent in Chapter 3. If V is a vector space, the set of linear maps taking V to itself will be denoted $\mathfrak{gl}(V)$.

Exercise 1.1.7 Consider $G = \mathbb{R}^+ \times \mathbb{R}$, with group product

$$\mu_G((x, a), (y, b)) = (xy, a + b)$$

and the group

$$H = \left\{ \begin{pmatrix} x & a \\ 0 & 1 \end{pmatrix} \mid x \in \mathbb{R}^+, a \in \mathbb{R} \right\},$$

with μ_H being matrix multiplication. Show that both G and H are groups. Although as *sets* G and H are in one-to-one correspondence, show that they are not isomorphic as groups. Hint: first show that if two groups are isomorphic and one of them is commutative, then so is the other.

Definition 1.1.8 For a square matrix $A = (a_{ij})$, we fix the following notations. The *determinant* of A is denoted $\det(A)$. The *complex conjugate* of A is $\bar{A} = (\overline{a_{ij}})$. The *transpose* of A is $A^T = (a_{ji})$. A *Hermitian* matrix satisfies $\bar{A}^T = A$. A *symmetric* matrix satisfies $A^T = A$. The $n \times n$ identity matrix is denoted I_n.

Example 1.1.9 The *special linear group* is

$$SL(n, \mathbb{R}) = \{A \in M_n(\mathbb{R}) \mid \det(A) = 1\}. \tag{1.6}$$

The general element has n^2 real parameters satisfying one condition, so $SL(n, \mathbb{R})$ has *dimension* $n^2 - 1$. The condition $\det(A) = 1$, which is polynomial in the parameters, defines a smooth surface in the parameter space $\mathbb{R}^{n^2}$.

More generally, if S is any $n \times n$ real matrix, then the set

$$G(n, S) = \{A \in GL(n, \mathbb{R}) \mid A^T SA = S\} \tag{1.7}$$

is a Lie group. An example is the Lorentz group (Definition 1.1.2).

Exercise 1.1.10 Prove $G(n, S)$ is a group. Show by example that S need be neither invertible nor symmetric, although these are the usual examples. If in (1.7), the matrix S is the identity matrix, then the group is $O(n)$, the orthogonal group. Specifically,

$$\begin{aligned} O(n) &= \{A \in GL(n, \mathbb{R}) \mid A^T A = I_n\} \\ SO(n) &= \{A \in GL(n, \mathbb{R}) \mid A^T A = I_n, \det(A) = 1\}. \end{aligned} \tag{1.8}$$

Let K be the diagonal matrix such that $K_{1,1} = -1$ and $K_{j,j} = 1$ for $j > 1$. Prove $O(n) = SO(n) \cup K \cdot SO(n)$.

Similarly, if S is an $n \times n$ complex matrix, then the set

$$\{A \in GL(n, \mathbb{C}) \,|\, \bar{A}^T S A = S\} \tag{1.9}$$

is a Lie group.

Example 1.1.11 The *special unitary group* $SU(n, \mathbb{C})$ is the set of $n \times n$ matrices with complex components satisfying both

$$\bar{U}^T U = I_n, \qquad \det(U) = 1.$$

It can be shown that

$$SU(2, \mathbb{C}) = \left\{ \begin{pmatrix} \alpha & \beta \\ -\bar{\beta} & \bar{\alpha} \end{pmatrix} \,|\, \alpha, \beta \in \mathbb{C}, \; \alpha\bar{\alpha} + \beta\bar{\beta} = 1 \right\}.$$

In other words, the general element of $SU(2)$ depends on three real parameters: the condition $\alpha\bar{\alpha} + \beta\bar{\beta} = 1$ can be written as $\alpha_1^2 + \alpha_2^2 + \beta_1^2 + \beta_2^2 = 1$ where we have set $\alpha = \alpha_1 + i\alpha_2$ and $\beta = \beta_1 + i\beta_2$. Thus, in the four dimensional real parameter space with coordinates $(\alpha_1, \alpha_2, \beta_1, \beta_2)$, the group $SU(2, \mathbb{C})$ is the unit sphere.

Lie groups of the form (1.7) and (1.9) are smooth surfaces (r dimensional, where r is the number of independent parameters describing the group elements) when viewed as sets in parameter space.

Definition 1.1.12 (Working definition) A *Lie group* is a group, which as a set is a smooth surface [†] in $\mathbb{R}^N$ for some N. Moreover, as functions of the parameters describing the surface, the product map μ and the inverse map $g \mapsto g^{-1}$ are smooth.

The rigorous definition requires a Lie group to be a *manifold*, that is, locally Euclidean. We will discuss the rigorous definition in detail in Chapter 2.

Definition 1.1.13 A set $\mathcal{T}$ of invertible maps taking some space X to itself is a *transformation group*, with the group product being composition of mappings, if,

(i) for all $f, g \in \mathcal{T}$, $f \circ g \in \mathcal{T}$,
(ii) the identity map id : $X \to X$, $\mathrm{id}(x) = x$ for all $x \in X$, is in $\mathcal{T}$, and
(iii) if $f \in \mathcal{T}$ then its inverse $f^{-1} \in \mathcal{T}$.

[†] More technically, a submanifold.

The associative law holds automatically for composition of mappings, and thus does not need to be checked. Matrix groups are groups of linear transformations since matrix multiplication and composition of linear maps coincide. We will assume that elements of transformation groups are smooth and are parametrised by either real or complex numbers, in a smooth way. This means that they are smooth when considered as maps of two sets of parameters, namely the group parameters and the independent variables of M.

Sets of transformations defined only on open sets of some space X can fail to be a group strictly as defined above; there are then technical difficulties with both closure and associativity when domains and ranges do not match, as in the next example.

Example 1.1.14 Let x and y be real variables and consider the map

$$(x, y) \mapsto \left(\frac{x}{1 - \epsilon x}, \frac{y}{1 - \epsilon x} \right).$$

We assume that ϵ is sufficiently close to zero so that no zero denominators result in our local domain of interest; note the restriction on ϵ depends on the point being mapped. Although ϵ parametrises a transformation of (x, y) space, we think of $\epsilon \in \mathbb{R}$ itself as acting on (x, y) space, so that we write the transformation as

$$(x, y) \mapsto \epsilon * (x, y). \tag{1.10}$$

We have in this case that

$$\epsilon_2 * (\epsilon_1 * (x, y)) = (\epsilon_1 + \epsilon_2) * (x, y), \tag{1.11}$$

where again, we need to restrict domains so that no zero denominators appear.

A set of invertible mappings such as that given in Example 1.1.14 is called a *local Lie transformation group*. See Olver (1993), Section 1.2 for the full technical definition. For the set of mappings given in Example 1.1.14, the identity transformation is parametrised by $\epsilon = 0$ and the transformation inverse to $\epsilon *$ is $(-\epsilon) *$, but since the domain of each transformation is different, we need to weaken the definition of closure to 'the composition of any two elements in the group is in the group, on the domain where the composition is defined'.

1.2 Actions

Given a (local) Lie group, we will be studying their *actions*, that is, their *presentations* as a group of transformations of some given space M. The simplest actions are linear actions, and the theory of such actions is the same as the theory of *representations* of the group.

Definition 1.2.1 If M is a vector space, a *representation* of the group G is a map $R : G \to GL(M)$ such that

$$R(gh) = R(g)R(h).$$

Exercise 1.2.2 Show that if $e \in G$ is the identity element in G, then $R(e)$ is the identity matrix, and that $R(g)^{-1} = R(g^{-1})$. Hint: for any element $g \in G$, $ge = g$.

Remark 1.2.3 The word *representation* is restricted, by common convention, to presentations in the form of a matrix group, which act on M linearly. Thus for representations, M is assumed to be a vector space.

Exercise 1.2.4 Let $G = SL(2)$ and $M = \mathbb{R}^3$. Show that the map

$$\begin{pmatrix} a & b \\ c & d \end{pmatrix} \mapsto \begin{pmatrix} a^2 & ab & b^2 \\ 2ac & ad + bc & 2bd \\ c^2 & cd & d^2 \end{pmatrix}$$

where $ad - bc = 1$ is a representation of $SL(2)$ in $GL(M)$.

Definition 1.2.5 We say the group G acts on the space M if there is a map

$$\alpha : G \times M \to M, \tag{1.12}$$

satisfying either

$$\alpha(g_2, \alpha(g_1, z)) = \alpha(g_2 g_1, z), \tag{1.13}$$

or

$$\alpha(g_2, \alpha(g_1, z)) = \alpha(g_1 g_2, z). \tag{1.14}$$

Actions obeying (1.13) are called *left actions* while actions obeying (1.14) are called *right actions*.

Exercise 1.2.6 Show that if $\alpha(g, .)$ is a left action then $\alpha(g^{-1}, .)$ is a right action, and vice versa.

One can define similarly a *local Lie group* action where the restrictions on the domains are noted.

In order to distinguish left from right actions, we will use the following notation. A left action will be denoted as

$$* : G \times M \to M$$

and thus (1.13) becomes $gh * z = g * (h * z)$. A right action will be denoted as

$$\bullet : G \times M \to M$$

and thus (1.14) becomes $gh \bullet z = h \bullet (g \bullet z)$.

The image of a point under a general action is denoted variously as

$$g \cdot z = \tilde{z} = F(z, g). \tag{1.15}$$

The different notations are used to ease the exposition, depending on the context.

Solution to Exercise 1.2.6   Given a left action $g * z$, define $g \bullet z = g^{-1} * z$. Then

$$h \bullet (g \bullet z) = h^{-1} * (g^{-1} * z) = (h^{-1}g^{-1}) * z = (gh)^{-1} * z = (gh) \bullet z$$

showing $g \bullet z$ is a right action as required. The other case is similar.

It is not always obvious whether a given action is left or right.

Exercise 1.2.7   Is the local action of $SL(2)$ on $\mathbb{R}$ given by

$$x \mapsto g \cdot x = \frac{ax + b}{cx + d}, \qquad ad - bc = 1 \tag{1.16}$$

left or right? Show that

$$h \cdot \frac{ax + b}{cx + d} = \frac{a(h \cdot x) + b}{c(h \cdot x) + d}$$

implies a right action, while

$$h \cdot \frac{ax + b}{cx + d} = \frac{a_2 \frac{ax+b}{cx+d} + b_2}{c_2 \frac{ax+b}{cx+d} + d_2}$$

is a left action, where

$$h = \begin{pmatrix} a_2 & b_2 \\ c_2 & d_2 \end{pmatrix}, \qquad a_2 d_2 - b_2 c_2 = 1.$$

The answer of whether a right or left action is implied by (1.16), depends on the interpretation of the symbol x, whether it is viewed as a coordinate function on $\mathbb{R}$ or an element of $\mathbb{R}$ itself. But which is which? (See section 1.3.1.)

Blanket assumption   We will assume that the space M on which G acts is a smooth space and that the map defining the action, $(g, z) \mapsto \alpha(g, z)$, is also smooth in both g and z.

Remark 1.2.8   Right actions are often denoted by $m \mapsto m \cdot g$, particularly in algebra texts discussing permutation groups. One then has $(m \cdot g) \cdot h = m \cdot (gh)$. Since in this book we are mostly doing calculus and not abstract algebra, we stick with the notion of the group action being a function on M and thus write it on the left hand side of its argument.

Example 1.2.9 A group acts on itself by left and right multiplication. The equations (1.13) and (1.14) are then the associative law for the group product.

Exercise 1.2.10 Show that $G \times G \to G$ given by

$$(g, h) \mapsto g^{-1} h g$$

is an action of G on itself. This is called the 'adjoint' or conjugation action.

Definition 1.2.11 Two group actions $\alpha_i : G \times M \to M, i = 1, 2$ are *equivalent* if there exists a smooth invertible map $\phi : M \to M$ such that

$$\alpha_2(g, z) = \phi^{-1} \alpha_1(g, \phi(z))$$

for all $g \in G$.

Exercise 1.2.12 Let $f : \mathbb{R} \to \mathbb{R}$ be any invertible map, and define $\mu : \mathbb{R} \times \mathbb{R} \to \mathbb{R}$ given by $\mu(x, y) = f^{-1}(f(x) + f(y))$. Show $(\mathbb{R}, \mu)$ is a group and thus defines an action of $\mathbb{R}$ on itself. Clearly, this action is equivalent to addition. Generalise this by taking invertible maps $f : (a, b) \subset \mathbb{R} \to \mathbb{R}$. A large number of seemingly mysterious non-linear group products on subsets of $\mathbb{R}$ can be generated this way. By considering $f = \arctan$, show the product

$$x \cdot y = \frac{x + y}{1 - xy}$$

is equivalent to addition.

A matrix group in $GL(n, \mathbb{R})$ acts on the n dimensional vector spaces V as a left action

$$A * \mathbf{v} = A\mathbf{v}$$

or a right action

$$A \bullet \mathbf{v} = A^T \mathbf{v},$$

where $\mathbf{v}$ is given as an $n \times 1$ vector with respect to some fixed basis of V.

Exercise 1.2.13 Show the two right actions, $A \bullet \mathbf{v} = A^T \mathbf{v}$ and $A \bullet \mathbf{v} = A^{-1} \mathbf{v}$ are *not* equivalent in general. That is, show that there does not exist, in general, a matrix ϕ such that $A^{-1} = \phi^{-1} A^T \phi$ for all A. Hint: the matrix ϕ must be independent of A for equivalence to hold.

Our next example concerns non-linear actions of $SL(2)$ on the plane. We will use these actions in many examples in the rest of this book.

Example 1.2.14 There are three inequivalent local actions of $SL(2, \mathbb{C})$ on the plane $\mathbb{C}^2$ (Olver, 1995). If the coordinates of $\mathbb{C}^2$ are taken to be (x, y) and the generic group element is

$$g = \begin{pmatrix} a & b \\ c & d \end{pmatrix}, \qquad ad - bc = 1$$

then the actions are as follows:

action 1

$$\widetilde{x} = \frac{ax + b}{cx + d}, \qquad \widetilde{y} = y \qquad (1.17)$$

action 2

$$\widetilde{x} = \frac{ax + b}{cx + d}, \qquad \widetilde{y} = \frac{y}{(cx + d)^2} \qquad (1.18)$$

action 3

$$\widetilde{x} = \frac{ax + b}{cx + d}, \qquad \widetilde{y} = 6c(cx + d) + (cx + d)^2 y. \qquad (1.19)$$

Keeping track of where repeated compositions of these maps are and are not defined is tedious. Usually one introduces a new point, ∞, and extends the definition of the action on x as follows:

$$\widetilde{x} = \frac{ax + b}{cx + d}, \qquad x \neq -\frac{d}{c}$$

$$\widetilde{-d/c} = \infty$$

$$\widetilde{\infty} = \frac{a}{c}.$$

Exercise 1.2.15 Which of these three actions of $SL(2)$ is equivalent to the standard linear action,

$$\widetilde{x} = ax + by, \qquad \widetilde{y} = cx + dy,$$

at least on some open set of $\mathbb{C}^2$? Hint: consider the induced action on $(x/\sqrt{y}, 1/\sqrt{y})$.

1.2.1 Semi-direct products

Suppose G and H are two Lie groups such that G acts on H, that is, $g * h \in H$ for all $g \in G$ and $h \in H$, for some left action $*$, satisfying

$$
\begin{aligned}
g * (h_1 h_2) &= (g * h_1)(g * h_2) && \text{for all } g \in G, h_1, h_2 \in H \\
e_G * h &= h && \text{for all } h \in H
\end{aligned}
\qquad (1.20)
$$

where e_G is the identity element of G. In words, the left action of each $g \in G$ is a homomorphism of H and e_G* is the identity map on H.

Exercise 1.2.16 Show that each $g \in G$ acts as an isomorphism. Show by example that the second condition in (1.20) is necessary for this to be true.

Definition 1.2.17 Suppose the action of G on H satisfies (1.20). The *semi-direct product* $G \ltimes H$ is defined to be, as a set, $G \times H$, but with group product $\cdot_\ltimes$ given by

$$
(g_1, h_1) \cdot_\ltimes (g_2, h_2) = (g_1 g_2, h_1(g_1 * h_2)).
$$

Exercise 1.2.18 Prove that the semi-direct product is associative. What is the identity element of $G \ltimes H$ and the inverse of (g, h)? Hence prove $G \ltimes H$ is a group.

The usual example is where G is an $n \times n$ real matrix Lie group and $H = (\mathbb{R}^n, +)$, the group of $n \times 1$ column vectors under addition. There is then the standard left action of G on H and the semi-direct product is represented by

$$
G \ltimes \mathbb{R}^n \approx \left\{ \begin{pmatrix} A & \mathbf{v} \\ 0 & 1 \end{pmatrix} \mid A \in G, \mathbf{v} \in \mathbb{R}^n \right\}.
\qquad (1.21)
$$

Indeed we have

$$
\begin{pmatrix} A & \mathbf{v} \\ 0 & 1 \end{pmatrix} \begin{pmatrix} B & \mathbf{w} \\ 0 & 1 \end{pmatrix} = \begin{pmatrix} AB & \mathbf{v} + A\mathbf{w} \\ 0 & 1 \end{pmatrix}
$$

as required.

Example 1.2.19 Recall the definition of the special orthogonal group $SO(n)$ given in Exercise 1.1.10. The semi-direct product $SO(n) \ltimes \mathbb{R}^n$ is called the *special Euclidean* group, denoted $SE(n)$, and is the transformation group generated by rotations and translations. Similarly, $SA(n) = SL(n) \ltimes \mathbb{R}^n$ is called the *special affine* group.

1.3 New actions from old

Given an action of G on M, there are induced actions on products of M, the set of functions defined on M, the tangent space of M and hence the set of vector fields on M, and so forth. We will start with the simplest of these and work our way up.

1.3.1 Induced actions on functions

The set of smooth functions mapping M to $\mathbb{R}^N$ is denoted $C^\infty(M, \mathbb{R}^N)$. A left action $G \times M \to M$ induces a right action on $C^\infty(M, \mathbb{R}^N)$ given by

$$g \bullet (f_1(z), \ldots, f_N(z)) = (f_1(g * z), \ldots, f_N(g * z)). \qquad (1.22)$$

This is a right action because

$$h \bullet (g \bullet f(z)) = h \bullet f(g * z) = f(g * (h * z)) = f(gh * z) = gh \bullet f(z).$$

Similarly, an action is induced on functions defined only on restricted domains in M; the domain of $g \bullet f$ will be $g^{-1} * \mathrm{domain}(f)$.

In particular, since coordinates are functions mapping M to $\mathbb{R}$, we have that a left action on M becomes a right action on the coordinates x_i:

$$x_i(g * z) = g \bullet x_i(z).$$

The image of a coordinate function under the action is denoted variously as

$$g \bullet x_j = \widetilde{x}_j = F_j(z, g), \qquad (1.23)$$

compare (1.15).

Consider the transformation group given in Example 1.1.14. The induced action on functions is given by

$$\epsilon \bullet f(x, y) = f(\epsilon * (x, y)) = f(\widetilde{x}, \widetilde{y}).$$

Thus

$$\epsilon \bullet \frac{x}{y} = \frac{\widetilde{x}}{\widetilde{y}} = \frac{x/(1 - \epsilon x)}{y/(1 - \epsilon x)} = \frac{x}{y}.$$

Definition 1.3.1 A function $f : M \to \mathbb{R}$ is said to be an *invariant* of the action $G \times M \to M$ if

$$f(g * z) = f(z), \qquad \text{for all } z \in M. \qquad (1.24)$$

Exercise 1.3.2 Consider the conjugation action of a real matrix group M on itself, given by $(A, B) \mapsto A^{-1}BA$. Show that the functions $\mathrm{tr}_n : M \to \mathbb{R}$ given by $\mathrm{tr}_n(B) = \mathrm{trace}(B^n)$, for $n \in \mathbb{N}$, are invariants for this action.

Exercise 1.3.3 The action on $C^\infty(M, \mathbb{R}^N)$ we have defined in equation (1.22) is an example of the following construction. Let G act on both M and N and let $\mathcal{M}(M, N)$ be the set of maps from M to N. If both these actions are left actions then there is an induced right action of G on $\mathcal{M}(M, N)$ given by

$$(g \cdot f)(x) = g^{-1} \cdot f(g \cdot x). \tag{1.25}$$

Show this is a right action. Show this is the same action on functions as described above in the case that $N = \mathbb{R}^n$ and $g \cdot n = n$ for all $g \in G, n \in N$. How should the action on $\mathcal{M}(M, N)$ be defined if G acts on the left on M and the right on N, to obtain a right action on $\mathcal{M}(M, N)$? In the case $N = M$, the action (1.25) is often called a *'gauge action'*.

1.3.2 Induced actions on products

Definition 1.3.4 The *product action* induced on the N-fold product of M with itself, $M \times \cdots \times M$ (N terms) is given by

$$g \cdot (z_1, \ldots, z_N) = (g \cdot z_1, \ldots, g \cdot z_N). \tag{1.26}$$

Example 1.3.5 The group generated by rotations and translations in $\mathbb{R}^n$ is known as the *Euclidean group* and is denoted $E(n)$. If $n = 2$, the standard action is

$$g_{(\theta,a,b)} * \begin{pmatrix} x \\ y \end{pmatrix} = \begin{pmatrix} \cos\theta & -\sin\theta \\ \sin\theta & \cos\theta \end{pmatrix} \begin{pmatrix} x \\ y \end{pmatrix} + \begin{pmatrix} a \\ b \end{pmatrix}. \tag{1.27}$$

The product action on two or more copies of $\mathbb{R}^2$ amounts to considering the simultaneous action on two or more points in the plane. Two invariants of the product action are

$$I_{n,m} = (x_n - x_m)^2 + (y_n - y_m)^2, \qquad J_{n,m} = x_n y_m - x_m y_n \tag{1.28}$$

where $(x_n, y_n), (x_m, y_m) \in \mathbb{R}^2$ and in fact any invariant of this group action is a function of the $I_{n,m}$ and the $J_{n,m}$. We will prove this in Section 4.7.

Definition 1.3.6 An N-*point invariant* of the action $G \times M \to M$ is an invariant of the product action on the N-fold product of M with itself.

These invariants are also known as *joint invariants*.

Exercise 1.3.7 Recall the local projective action of $SL(2)$ on $\mathbb{R}$,

$$x \mapsto (ax + b)/(cx + d) = \widetilde{x},$$

where $ad - bc = 1$. Take the product action on (x, y, z, w) space, that is,

$$(x, y, z, w) \mapsto (\widetilde{x}, \widetilde{y}, \widetilde{z}, \widetilde{w}).$$

Show that the *cross ratio*

$$\frac{(x - z)(y - w)}{(x - y)(z - w)}$$

is a 4-point invariant.

1.3.3 Induced actions on curves

If G acts on M, s is a real parameter and $s \mapsto \gamma(s) \in M$ is thus a curve in M, the induced action on the curve is defined pointwise,

$$(g * \gamma)(s) = g * \gamma(s),$$

see Figure 1.1. Since the action is smooth and invertible, it will not introduce cusps or self-crossings into curves that do not have them to begin with. As simple as this looks, it is probably one of the most important induced actions in this book because the applications are so widespread; the curve might be a solution curve of a differential equation, it might be a path of a particle in some physical system or a light ray in an optical medium, it might be a 'tangent element', and so on.

Exercise 1.3.8 Show a matrix group acting linearly on a vector space V, on the left, induces an action on the set of lines passing through the origin of V. If the dimension of V is 2, show the induced action on the slope m of a line is

$$\widetilde{m} = \begin{pmatrix} a & b \\ c & d \end{pmatrix} \bullet m = (c + dm)/(a + bm).$$

Hint: consider the induced action on y/x.

If the curve is differentiable, we obtain an induced action on the first and higher order derivatives, called the prolonged action, discussed in the next section 1.3.4.

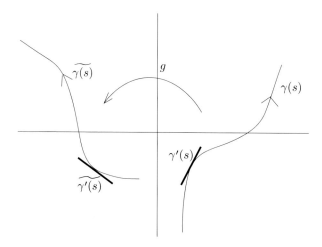

Figure 1.1 Euclidean action on smooth curves in the plane.

1.3.4 Induced action on derivatives: the prolonged action

Suppose there is an action of the group G in the plane with coordinates (x, y). If we take a curve in the plane given by $y = f(x)$, so that we consider y to be a function of x, then there is an induced action on the derivatives y_x, y_{xx} and so forth, called the *prolonged* action.

To illustrate, we look in detail at the action in Example 1.3.5. To ease the notation, we denote the generic image of this action as

$$g_{(\theta,a,b)} * \begin{pmatrix} x \\ y \end{pmatrix} = \begin{pmatrix} \widetilde{x} \\ \widetilde{y} \end{pmatrix}$$

$$g_{(\theta,a,b)} \cdot \frac{\mathrm{d}y}{\mathrm{d}x} = \widetilde{y}_x \qquad (1.29)$$

$$g_{(\theta,a,b)} \cdot \frac{\mathrm{d}^2 y}{\mathrm{d}x^2} = \widetilde{y}_{xx}$$

and define the action on y_x to be

$$g_{(\theta,a,b)} \cdot \frac{\mathrm{d}y}{\mathrm{d}x} = \frac{\mathrm{d}\widetilde{y}}{\mathrm{d}\widetilde{x}} = \frac{\mathrm{d}\widetilde{y}/\mathrm{d}x}{\mathrm{d}\widetilde{x}/\mathrm{d}x}, \qquad (1.30)$$

by the chain rule (which is, in one dimension, also known as implicit differentiation). Thus

$$\widetilde{y}_x = \frac{\sin\theta + \cos\theta\, y_x}{\cos\theta - \sin\theta\, y_x}.$$

Similarly,

$$g_{(\theta,a,b)} \cdot \frac{\mathrm{d}^2 y}{\mathrm{d}x^2} = \frac{\mathrm{d}^2 \widetilde{y}}{\mathrm{d}(\widetilde{x})^2} = \widetilde{y}_{\widetilde{x}\widetilde{x}} = \frac{1}{\mathrm{d}\widetilde{x}/\mathrm{d}x} \frac{\mathrm{d}}{\mathrm{d}x} \frac{\mathrm{d}\widetilde{y}/\mathrm{d}x}{\mathrm{d}\widetilde{x}/\mathrm{d}x}$$

so we have

$$\widetilde{y_{xx}} = \frac{y_{xx}}{(\cos\theta - \sin\theta y_x)^3}.$$

Since

$$\frac{\widetilde{y_{xx}}}{(1 + \widetilde{y_x}^2)^{3/2}} = \frac{y_{xx}}{(1 + y_x^2)^{3/2}} \tag{1.31}$$

we say that $y_{xx}/(1 + y_x^2)^{3/2}$ is a *differential invariant* (see Definition 1.3.11). This quantity is in fact the Euclidean curvature of the path $x \mapsto (x, y(x))$ in the plane.

Exercise 1.3.9 For the action of $SL(2)$ on (x, u) space given by

$$\widetilde{x} = x, \qquad \widetilde{u} = \frac{au + b}{cu + d}, \qquad ad - bc = 1$$

show that

$$\frac{u_{xxx}}{u_x} - \frac{3}{2} \frac{u_{xx}^2}{u_x^2} \tag{1.32}$$

is invariant under the prolonged action. This expression is known as the *Schwarzian derivative* of u with respect to x and is often denoted $\{u; x\}$.

More generally, we are concerned with q smooth functions u^α that depend on p variables x_i. The derivatives of these will be denoted using a multi-index notation, e.g.

$$u^\alpha_{1112222} = \frac{\partial^7}{\partial x_1^3 \partial x_2^4} u^\alpha$$

or

$$u^\beta_{xxyyy} = \frac{\partial^5}{\partial x^2 \partial y^3} u^\beta.$$

We consider these derivative functions as functionally independent coordinates of a so-called *jet space*, denoted $\mathcal{J}(X \times U)$, or $\mathcal{J}$ for short, where X is the space whose coordinates are the independent variables, and U the space whose coordinates are the dependent variables. A differential equation is thus a surface in $\mathcal{J}$. If we restrict the order of the derivative to be n, we denote the resulting space by $\mathcal{J}^n(X \times U)$. Points in $\mathcal{J}(X \times U)$ have coordinates

$$z = z(x_1, \dots, x_p, u^1, \dots, u^q, u_1^1, \dots).$$

Readers interested in the rigorous formulation of jet spaces as fibre bundles should consult, for example, Saunders (1989). However, it is not necessary to know the considerable technical details that are involved in order to apply group actions to differential equations effectively. Indeed, a solid grasp of the multivariable chain rule is far more important.

Our interest is in functions, like the Schwarzian above, that depend on a finite number of derivatives and are smoothly differentiable with respect to those arguments (away from any zeros in denominators).

Definition 1.3.10 We will denote by $\mathcal{A}$ the algebra of smooth functions on $\mathcal{J}(X \times U)$, that depend on finitely many arguments.

The operator $\partial/\partial x_i$ extends to an operator on $\mathcal{A}$ called the *total differentiation operator*

$$D_i = \frac{D}{Dx_i} = \frac{\partial}{\partial x_i} + \sum_{\alpha=1}^{q} \sum_{K} u_{Ki}^{\alpha} \frac{\partial}{\partial u_K^{\alpha}}. \tag{1.33}$$

We assume we are given a smooth left action of an r dimensional Lie group G on the space $X \times U$, where X is the space of independent variables and U is the space of dependent variables. By prolongation we will get a right action on the derivatives u_K^{α}, where K is the multi-index of differentiation, which is calculated using the chain rule of differentiation; a right action since the u_K^{α} are coordinates of the relevant jet bundle. The prolonged action will be denoted variously as

$$g \bullet u_K^{\alpha} = \widetilde{u_K^{\alpha}} = F_K^{\alpha}(z, g),$$

compare (1.15) and (1.23). This right action then extends to an action on $\mathcal{A}$, as in Section 1.3.1.

Definition 1.3.11 Given a smooth action $G \times X \times U \to X \times U$, a *differential invariant* is an element of $\mathcal{A}$ which is invariant under the induced prolonged action.

The prolonged action is given explicitly by

$$g \bullet u_{i..j}^{\alpha} = \widetilde{D}_i \cdots \widetilde{D}_j F^{\alpha}(z, g), \tag{1.34}$$

where

$$\widetilde{D}_i = \frac{D}{D\widetilde{x}_i} = \sum_{k=1}^{p} (\widetilde{D}x)_{ik} D_k \tag{1.35}$$

and the coefficients are obtained from the Jacobian matrix of the coordinate transformation $x \mapsto \widetilde{x}$,

$$(\widetilde{D}x)_{ik} = ((D\widetilde{x})^{-1})_{ik}.$$

Explicitly, we have

$$D\widetilde{x} = \begin{pmatrix} \dfrac{\partial \widetilde{x}_1}{\partial x_1} & \cdots & \dfrac{\partial \widetilde{x}_1}{\partial x_p} \\ \vdots & \ddots & \vdots \\ \dfrac{\partial \widetilde{x}_p}{\partial x_1} & \cdots & \dfrac{\partial \widetilde{x}_p}{\partial x_p} \end{pmatrix} \tag{1.36}$$

and note the fact that the Jacobian of the inverse map is the inverse of the Jacobian.

Example 1.3.12 Consider the action of the Euclidean group $SE(2)$ on the (u, v) plane given by,

$$\begin{pmatrix} \widetilde{u} \\ \widetilde{v} \end{pmatrix} = \begin{pmatrix} \cos\theta & -\sin\theta \\ \sin\theta & \cos\theta \end{pmatrix} \begin{pmatrix} u \\ v \end{pmatrix} + \begin{pmatrix} a \\ b \end{pmatrix}.$$

Assume $u = u(x, t)$ and $v = v(x, t)$. Since x and t are *both* invariant, we have $\widetilde{D}_x = D_x$, $\widetilde{D}_t = D_t$ and hence the prolonged action is

$$g \bullet \begin{pmatrix} u_K \\ v_K \end{pmatrix} = \begin{pmatrix} \cos\theta & -\sin\theta \\ \sin\theta & \cos\theta \end{pmatrix} \begin{pmatrix} u_K \\ v_K \end{pmatrix}.$$

Example 1.3.13 For the group $SL(2)$ acting on the variables $(x, t, u(x, t))$ as $\widetilde{t} = t$ and

$$\begin{pmatrix} \widetilde{x} \\ \widetilde{u} \end{pmatrix} = \begin{pmatrix} a & b \\ c & (1 + bc)/a \end{pmatrix} \begin{pmatrix} x \\ u \end{pmatrix} \tag{1.37}$$

where (a, b, c) are the coordinates of $g \in SL(2)$ near the identity $e = (1, 0, 0)$, we have courtesy of the chain rule that

$$\begin{pmatrix} D_x \\ D_t \end{pmatrix} = \begin{pmatrix} a + bu_x & 0 \\ bu_t & 1 \end{pmatrix} \begin{pmatrix} \widetilde{D}_x \\ \widetilde{D}_t \end{pmatrix}$$

and thus

$$\widetilde{D}_x = \frac{1}{a + bu_x} D_x, \qquad \widetilde{D}_t = D_t - \frac{bu_t}{a + bu_x} D_x.$$

Note that even though $\widetilde{t} = t$, it is *not* the case that $\widetilde{D}_t = D_t$. From equation (1.34) it now follows that

$$\widetilde{u}_x = \widetilde{D}_x \widetilde{u}, \qquad \widetilde{u}_{xx} = \widetilde{D}_x^2 \widetilde{u}, \qquad \widetilde{u}_t = \widetilde{D}_t \widetilde{u},$$

which yields

$$\widetilde{u}_x = \frac{ac + u_x(1 + bc)}{a(a + bu_x)}, \qquad \widetilde{u_{xx}} = \frac{u_{xx}}{(a + bu_x)^3}, \qquad \widetilde{u_t} = \frac{u_t}{a + bu_x}.$$

It can be checked that this is a right action.

1.3.5 Some typical group actions in geometry and algebra

By and large, group actions in geometry and algebra are induced from linear actions on vector spaces.

If G is a matrix Lie group in $GL(n)$ and V is an n dimensional vector space with basis $e_1, \ldots, e_n$, then there is the standard left action of G on V with respect to the given basis given by

$$A \cdot e_i = \sum_{j=1}^n a_{ij} e_j, \qquad A = (a_{ij}) \in G \qquad (1.38)$$

and extended linearly.

We have already seen the induced action on products, and the induced left action on $V \times V \times \cdots \times V$ is just that. The induced left action on the tensor product, $V \otimes V$ which has basis $\{e_i \otimes e_j\}$ is given by

$$\widetilde{e_i \otimes e_j} = \widetilde{e}_i \otimes \widetilde{e}_j,$$

and extended linearly. Thus if A is a matrix in the group acting on V, with matrix form (a_{ij}), then there is a listing of the basis elements of $V \otimes V$ such that the matrix form of $A \otimes A$ is given by

$$\begin{pmatrix} a_{11}A & \cdots & a_{1n}A \\ \vdots & \ddots & \vdots \\ a_{n1}A & \cdots & a_{nn}A \end{pmatrix}.$$

Restricting ourselves to the space of symmetric tensors, $S^2(V)$, with basis $\{\frac{1}{2}(e_i \otimes e_j + e_j \otimes e_i)\}$, or antisymmetric tensors, $\Lambda^2(V)$, with basis $\{\frac{1}{2}(e_i \otimes e_j - e_j \otimes e_i) \mid i \neq j\}$, the induced action on $V \otimes V$ given above takes these subspaces to themselves and thus we obtain actions on these subspaces. Similar remarks apply to spaces of n-fold symmetric and antisymmetric tensors, denoted $S^n(V)$ and $\Lambda^n(V)$.

Remark 1.3.14 The set of n-fold symmetric tensors, $S^n(V)$, should not be confused with the unit sphere in V in the case $n = \dim(V)$.

Similarly, given linear actions by the same group, but on two different vector spaces V and W, we obtain induced actions on $V \times W$ and $V \otimes W$.

More interesting is the induced action on the dual V^* of V. The simplest way to think of the dual is as the space of coefficients $(a_1, a_2, \ldots, a_n)$ of a generic element of V, $v = a_1 e_1 + a_2 e_2 + \cdots + a_n e_n$. The group acts on this element as

$$\tilde{v} = a_1 \tilde{e}_1 + a_2 \tilde{e}_2 + \cdots + a_n \tilde{e}_n.$$

Expanding out $\tilde{e}_i$ using equation (1.38) above, we obtain by collecting terms,

$$\tilde{v} = \tilde{a}_1 e_1 + \tilde{a}_2 e_2 + \cdots + \tilde{a}_n e_n.$$

Then $\mathbf{a} = (a_1, \ldots, a_n) \mapsto \tilde{\mathbf{a}} = (\tilde{a}_1, \ldots, \tilde{a}_n)$ is a right action.

Exercise 1.3.15 Show that if g has matrix A with respect to the basis e_i, $i = 1, \ldots, n$, so that $\tilde{e}_i = \sum_j A_{ij} e_j$, then $\tilde{\mathbf{a}} = \mathbf{a}A$.

Similarly, we have actions induced on the dual of $S^n(V)$. A typical element in $S^n(V)$ is written as a symbolic polynomial in the e_i; since the products are symmetric, this makes sense. Applying the action to the e_i, expanding and collecting coefficients leads to an action on the coefficients, and hence on the dual of $S^n(V)$. One of the most important examples of this construction, at least historically for a physicist, is the induced actions of $SU(2)$ on the coefficients of a generic homogeneous polynomial of degree 2 and above. We refer the reader to Fässler and Stiefel (1992), Section 4.3.1, where the details are given in full. It is well worth the student's while, from a general educational point of view, to understand the details of the calculations not only of this induced group action but also those of the Clebsch–Gordan Theorem, which gives the decomposition into direct summands of a tensor product of these actions.

Other types of actions induced from linear actions in vector spaces are those on sets of lines or planes passing through the origin in V.

Example 1.3.16 Consider the set of straight lines passing through the origin in $\mathbb{R}^{n+1}$. Then any linear map sends one line to another. Such a line intersects the unit sphere in two antipodal points. The set of pairs of antipodal points on an n-sphere is called the n dimensional projective space, $\mathbb{P}^n$. Thus a group of matrices acting in $\mathbb{R}^{n+1}$ induces an action on $\mathbb{P}^n$. This space is not linear and so neither is the induced action.

Exercise 1.3.17 Just as groups of matrices induce actions on straight lines through the origin, so they induce actions on sets of planes through the origin.

Consider the $(2n) \times (2n)$ matrix,

$$J = \begin{pmatrix} 0 & I_n \\ -I_n & 0 \end{pmatrix}$$

where I_n is the $n \times n$ identity matrix and 0 the $n \times n$ zero matrix. Given a subspace W of $\mathbb{R}^{2n}$, define

$$W^{\perp} = \{v \in \mathbb{R}^{2n} \mid v^T J w = 0, \text{ for all } w \in W\}.$$

If $W = W^{\perp}$ then W is called a *Lagrangian subspace*. Show that the dimension of any Lagrangian space is n. The *symplectic group* $Sp(n)$ is the set of $2n \times 2n$ matrices A such that

$$A^T J A = J.$$

Show that $Sp(n)$ is a group, and that it maps the set of Lagrangian spaces to itself, that is, if $W = W^{\perp}$ and $A \in Sp(n)$, then $AW = (AW)^{\perp}$.

1.4 Properties of actions

Not all group actions are of interest. To define a moving frame, the main subject of this book, we need an action to be *free* and *regular* at least in the neighbourhood of our space where we want the frame. Freeness means that if $g \cdot z = z$, then $g = e$, the identity element. Regularity relates to how the group orbits 'foliate' the space.

Definition 1.4.1 Let G act on M and let $z \in M$. The *orbit* of z is the set of points in M that are the image of z under the group action,

$$\mathcal{O}(z) = \{g \cdot z \mid g \in G\}.$$

If we write the space M as a union of orbits of a Lie group action, we have what is known as a *foliation* of M, with each orbit being a *leaf* of the foliation. In general, the leaves will not all have the same dimension, see Figure 1.2.

A *regular* foliation of an n dimensional space has the property that there exists a local coordinate transformation and an integer r such that the leaves are mapped to the set of planes

$$\{(k_1, k_2, \ldots, k_{n-r-1}, z_{n-r}, \ldots, z_n\},$$

where the k_i are constants. For example, if the group action is translation in the last r coordinate directions, the orbits are already planes. For the action

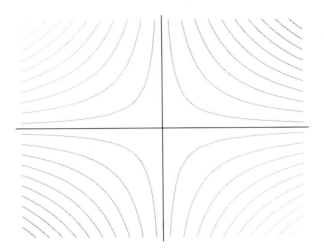

Figure 1.2 The orbits of the action $\widetilde{x} = \exp(\alpha)x$, $\widetilde{y} = \exp(-\alpha)y$ foliate the plane. The origin is a fixed point and the foliation is not regular there.

shown in Figure 1.2, a foliation map exists in a disc around any point in $\mathbb{R}^2$ except the origin, that is, there is a coordinate transformation that straightens out the orbits. If the orbits all have the same dimension, then the foliation they yield is said to be regular. More is required for an action to be regular however.

Most actions are not free and regular, but they can often be extended in various ways so that they become free and regular, at least for worthwhile portions of the extended space. In particular, this is true for prolongation actions, provided the action you start with is *locally effective on subsets*. Here we look in detail at these concepts.

The first notion we define is the isotropy group of a subset S of M. The notion of an isotropy group is 'dual' to that of a fixed point or invariant set. It answers the question, given a set, which elements of the group fix that set? Note that not every element of S is a fixed point of the isotropy group, but it needs to map to another element of S.

Definition 1.4.2 [Isotropy groups] For $S \subset M$ define the *isotropy group* of S to be $G_S = \{g \in G \mid g \cdot S \subset S\} = \{g \in G \mid g \cdot z \in S, \forall z \in S\}$. The *global isotropy group* of S is defined to be

$$G_S^* = \cap_{z \in S} G_z = \{g \in G \mid g \cdot z = z, \forall z \in S\}.$$

Exercise 1.4.3 For the group of translations and rotations in the plane, what is the isotropy group of an arbitrary given point? (Hint: rotate your point and then translate it back to itself.) What if that point is the origin? What is the isotropy group of a circle in the plane? What is the global isotropy group of a circle in the plane?

Definition 1.4.4 A *discrete subgroup* of a Lie group G is a subgroup which, as a set, consists of isolated points in G.

Definition 1.4.5 (Free and effective actions) A group action on M is said to be:

free	if	$G_z = \{e\}$, for all $z \in M$
locally free	if	G_z is a discrete subgroup of G, for all $z \in M$
effective	if	$G_M^* = \{e\}$
locally effective	if	G_U^* is a discrete subgroup of G
on subsets		for every open subset $U \subset M$.

Exercise 1.4.6 Show the standard action of the Euclidean group on the plane is effective but not free. If you induce the action on curves and prolong, at what degree of prolongation does the action become free?

Remark 1.4.7 A theorem of Ovsiannikov (Ovsiannikov, 1982, see also Olver, 2000) guarantees that the prolongation of actions, which are locally effective on subsets of $\mathcal{X} \times \mathcal{U}$, will be locally free on an open dense subset of $J^n(\mathcal{X} \times \mathcal{U})$ for sufficiently large n. For results for product actions, see Boutin (2002) and Olver (2001a), Example 3.1.

Definition 1.4.8 A group action is *regular* if

(i) all orbits have the same dimension,
(ii) for each $z \in M$, there are arbitrarily small neighbourhoods $\mathcal{U}(z)$ of z such that for all $z' \in \mathcal{U}(z), \mathcal{U}(z) \cap \mathcal{O}(z')$ is connected.

Locally, a regular action has orbits that foliate the space in a regular way, with one additional condition (ii) above, see Figure 1.3.

An example of an action satisfying (i) but not (ii) in Definition 1.4.8 is pictured in Figure 1.4. The group is $(\mathbb{R}, +)$ and acting on the punctured plane, $\mathbb{R}^2 \setminus \{(0, 0)\}$; the orbits are the integral curves of the flow shown. The action is not regular at those points lying on the periodic orbit P. Indeed, for every $z \in P$, the intersection of arbitrarily small neighbourhoods $\mathcal{U}(z)$ with orbits of points $z' \notin P$ have infinitely many components.

The next notion we require is that of the *transversality* of surfaces in our space M. Figure 1.5 illustrates the concept for curves in the plane.

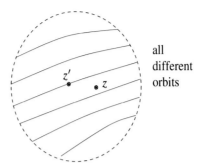

Figure 1.3 A regular action foliates the space with orbits, all of the same dimension. Moreover, each orbit intersects the neighbourhood in a single connected component.

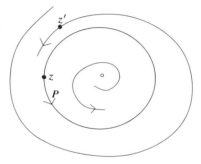

Figure 1.4 An action which is not regular at points $z \in P$ where P is the periodic orbit.

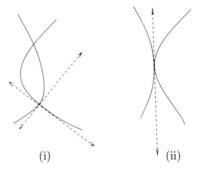

(i)   (ii)

Figure 1.5 The two curves at (i) intersect transversally as at every intersection point the two tangent spaces of the curves span the tangent space of the plane. The two curves at (ii) are not transverse as the span of the tangent spaces at the intersection has dimension less than that of the ambient space.

Definition 1.4.9 We say two smooth surfaces $\mathcal{K}$ and $\mathcal{O}$ contained in $\mathbb{R}^n$, of dimensions α and β respectively, $0 \leq \alpha, \beta \leq n, \alpha + \beta \geq n$, intersect *transversally* if for every $z \in \mathcal{K} \cap \mathcal{O}$, the tangent spaces $T_z\mathcal{K}$ and $T_z\mathcal{O}$, viewed as subspaces of $T_z\mathbb{R}^n$, satisfy

$$T_x\mathcal{K} + T_x\mathcal{O} = T_x\mathbb{R}^n;$$

in words, the span of the two tangent spaces is the full tangent space at every point of intersection.

If the action is regular, then locally there exists a smooth cross section $\mathcal{K}$, see Figure 4.1, such that for the orbits $\mathcal{O}(z)$,

(i) $\dim \mathcal{K} + \dim \mathcal{O}(z) = \dim M$,
(ii) for $z \in \mathcal{K}$, $\mathcal{K} \cap \mathcal{O}(z)$ is a single point,
(iii) $\mathcal{K}$ is transverse to the group orbits; that is, for $z \in \mathcal{K}$, the direct sum of the tangent spaces of $\mathcal{K}$ and $\mathcal{O}(z)$ at the point z is the whole of $T_z M$.

As we will show in Chapter 4, the cross section $\mathcal{K}$ *is* the moving frame that is the subject of this book.

1.5 One parameter Lie groups

The easiest way to analyse a group action in detail is to examine the action of its one parameter subgroups.

Definition 1.5.1 A *one parameter Lie subgroup* of a Lie group G is a path $t \mapsto h(t) \in G$ such that

$$\begin{aligned} h(0) &= e \\ h(t)h(s) &= h(t + s), \qquad \text{for all } s, t \in \mathbb{R} \end{aligned} \tag{1.39}$$

Example 1.5.2 In $G = \mathbb{C} \setminus \{0\}$, with the usual product, let $h(t) = \mathrm{e}^{\mathrm{i}t}$. This is a one parameter subgroup as $h(t)h(s) = \mathrm{e}^{\mathrm{i}t}\mathrm{e}^{\mathrm{i}s} = \mathrm{e}^{\mathrm{i}(t+s)}$, and $h(0) = 1$.

Exercise 1.5.3 Show that if $\alpha \in \mathbb{C}$ is non-zero, then $h(t) = \mathrm{e}^{\alpha t}$ is a one parameter subgroup of $\mathbb{C} \setminus \{0\}$. Plot the curves in $\mathbb{C}$ for various α.

Exercise 1.5.4 Let $G = \mathbb{R}^+ \times \mathbb{R}$ with group product given by

$$(c, d) \cdot (a, b) = (ac, cb + d).$$

Show $h_1(t) = (1, \beta t)$ and $h_2(t) = (\mathrm{e}^{\alpha t}, 0)$ are both one parameter subgroups. For arbitrary $\alpha, \beta \in \mathbb{R}$, $\alpha \neq 0$, show the general form of the one parameter

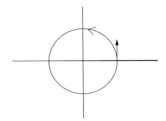

Figure 1.6 The one parameter group $\{\exp(it) \mid t \in \mathbb{R}\}$ in the complex plane.

subgroups is

$$h_3(t) = (e^{\alpha t}, \beta(e^{\alpha t} - 1)/\alpha).$$

To each one parameter subgroup $h(t)$ we can associate a vector $\mathbf{v}_h$ at the identity element of G, the tangent vector to path at $t = 0$,

$$\mathbf{v}_h = \left.\frac{\mathrm{d}}{\mathrm{d}t}\right|_{t=0} h(t), \qquad (1.40)$$

see Figure 1.6. In Chapter 2, we will clarify and prove the important theorem,

Theorem 1.5.5 *One parameter subgroups of a group G are in one-to-one correspondence with tangent vectors at the identity element of G.*

In interpreting this theorem, we distinguish $h(t)$ from, say, $h(2t)$, even though the images of the curves are the same. Unless otherwise stated, the parametrisation is an intrinsic component of a one parameter subgroup. Different parametrisations can be distinguished by their value at $t = 1$.

For matrix groups, tangent vectors of one parameter subgroups can be easily computed. Indeed, if $A(t) = (a_{ij}(t))$, then $A'(t)$ is the matrix $A'(t) = (a'_{ij}(t))$.

Example 1.5.6 Let $G = O(3) = \{A \in GL(3, \mathbb{R}) : A^T A = I\}$, that is, the group of 3×3 orthogonal matrices. Let the one parameter subgroup $h(t)$ be given by

$$h(t) = \begin{pmatrix} \cos t & -\sin t & 0 \\ \sin t & \cos t & 0 \\ 0 & 0 & 1 \end{pmatrix}.$$

Then the associated tangent vector is

$$\mathbf{v}_h = \left.\frac{\mathrm{d}}{\mathrm{d}t}\right|_{t=0} h(t) = \begin{pmatrix} 0 & -1 & 0 \\ 1 & 0 & 0 \\ 0 & 0 & 0 \end{pmatrix}.$$

Exercise 1.5.7 Show that

$$
h(t) = \begin{pmatrix} \cosh(\mu t) + \dfrac{\alpha}{\mu}\sinh(\mu t) & \dfrac{\beta}{\mu}\sinh(\mu t) \\[2ex] \dfrac{\gamma}{\mu}\sinh(\mu t) & \cosh(\mu t) - \dfrac{\alpha}{\mu}\sinh(\mu t) \end{pmatrix}
$$

is a one parameter subgroup of $SL(2)$, that is, not only $h(s)h(t) = h(s+t)$ but also $\det h(t) = 1$, provided $\mu^2 = \alpha^2 + \beta\gamma$. Show

$$
\mathbf{v}_h = \begin{pmatrix} \alpha & \beta \\ \gamma & -\alpha \end{pmatrix}.
$$

Show that any tangent vector at the identity of $SL(2)$ has this form, that is, has zero trace. Hint: differentiate $\det(A(t)) - 1 = 0$ with $A(0) = I_2$.

1.6 The infinitesimal vector fields

A rigorous discussion of the ideas in this section requires the concepts discussed in Chapters 2 and 3. Fortunately, for many applications the calculation of derivatives and tangent vectors is well defined from the context. The informal discussion here is for these cases.

Definition 1.6.1 If $h(t)$ is a one parameter subgroup of G acting on M, so that $\mathbf{v}_h$ is as defined in equation (1.40), and supposing that differentiation on M is defined, then the *infinitesimal action* of $h(t)$ at $z \in M$ is the *vector*

$$
\mathbf{v}_h \cdot z = \left. \frac{d}{dt} \right|_{t=0} h(t) \cdot z.
$$

The *infinitesimal vector field* is the map,

$$
z \mapsto \mathbf{v}_h \cdot z,
$$

giving a vector at every point z (see Figure 1.7).

Note that 'the infinitesimal action' is *not* a group action; rather the vector fields represent the associated Lie algebra, which is defined in Chapter 3.

Exercise 1.6.2 For a one parameter matrix group $h(t)$ acting linearly on the left (right) of a vector space V, show the infinitesimal action is simply left (right) multiplication by the matrix $\mathbf{v}_h$. Hint: the product rule holds for the matrices.

Exercise 1.6.3 For a one parameter matrix subgroup $h(t) \subset G$ acting by left (right) multiplication on G, show the infinitesimal action is simply left (right) multiplication by $\mathbf{v}_h$.

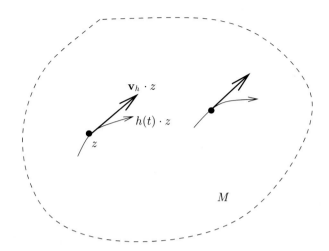

Figure 1.7 The infinitesimal action of a one parameter Lie group gives a vector field on M.

Example 1.6.4 Given a matrix Lie group G, the *adjoint* or *conjugation* action of a one parameter Lie subgroup $t \mapsto H(t) \in G$ on all of G is

$$\widetilde{A}(t) = H(t)^{-1} A H(t).$$

Since $H(t)^{-1} H(t) \equiv I$ and $H(0) = I$, we have that $(H^{-1})'(0) = -H'(0) = -\mathbf{v}_H$, and so the infinitesimal action is

$$\left.\frac{\mathrm{d}}{\mathrm{d}t}\right|_{t=0} \widetilde{A}(t) = A\mathbf{v}_H - \mathbf{v}_H A = [A, \mathbf{v}_H].$$

If near $z \in M$ we have coordinates $z = (z_1, \ldots, z_n)$, then $h(t) \cdot z = (\widetilde{z}_1(t), \widetilde{z}_2(t), \ldots, \widetilde{z}_n(t))$ and the infinitesimal action is calculated component-wise. In the particular case of a prolonged action in $\mathcal{J}(X \times U)$, the infinitesimal action is calculated on the coordinates u_K^α as

$$\mathbf{v}_h \cdot u_K^\alpha = \left.\frac{\mathrm{d}}{\mathrm{d}t}\right|_{t=0} \widetilde{u_K^\alpha}(t).$$

Example 1.6.5 For the one parameter subgroup of $SL(2)$ given in Exercise 1.5.7, and the first of the actions of $SL(2)$ given in Example 1.2.14,

$$\widetilde{x} = \frac{ax + b}{cx + d}, \qquad \widetilde{y} = y, \qquad ad - bc = 1$$

we have that

$$\mathbf{v}_h \cdot (x, y, y_x) = (2\alpha x + \beta - \gamma x^2, 0, 2y_x(\gamma x - \alpha)).$$

The first two components follow directly from the group action as given, while the third component follows from the prolonged action which is

$$\widetilde{y}_x(t) = \frac{1}{\mu^2} y_x (\mu \cosh(\mu t) - (\alpha - \gamma x) \sinh(\mu t))^2$$

where we have used $\mu^2 = a^2 + bc$ to simplify the expression.

Exercise 1.6.6 Consider the one parameter subgroup of $SL(2)$ given in Exercise 1.5.7, and the third of the actions of $SL(2)$ given in Example 1.2.14. Calculate the infinitesimal action at (x, y, y_x).

Recall for a function f on M we defined the induced action on f to be $(g \cdot f)(z) = f(g \cdot z)$ (Section 1.3.1).

Definition 1.6.7 For a differentiable function f defined on M and a one parameter group $h(t)$ acting on M, we define the infinitesimal action on f to be

$$(\mathbf{v}_h \cdot f)(z) = \frac{\mathrm{d}}{\mathrm{d}t}\bigg|_{t=0} f(h(t) \cdot z).$$

Thus if $f : M \to \mathbb{R}$ and if in coordinates $z = (z_1, z_2, \ldots, z_n)$ then

$$(\mathbf{v}_h \cdot f)(z) = \sum_i \frac{\partial f}{\partial z_i} \mathbf{v}_h \cdot z_i. \tag{1.41}$$

Looking again at Example 1.6.5, for a given real valued function $f = f(x, y, y_x)$ we have

$$(\mathbf{v}_h \cdot f)(x, y, y_x) = (2\alpha x + \beta - \gamma x^2)\frac{\partial f}{\partial x} + 2y_x(\gamma x - \alpha)\frac{\partial f}{\partial y_x}.$$

Proposition 1.6.8 *If* $f : M \to \mathbb{R}$ *is an invariant of the group action* $G \times M \to M$, *then for every one parameter subgroup* $h(t) \subset G$, $\mathbf{v}_h \cdot f \equiv 0$.

Exercise 1.6.9 Prove Proposition 1.6.8. Hint: $f(h(t) \cdot z) \equiv f(z)$.

Definition 1.6.10 If $\mathbf{v}_h \cdot f \equiv 0$ for every one parameter subgroup $h(t) \subset G$, we say that f satisfies the *infinitesimal criterion for invariance*.

Let us look in detail at the infinitesimal action for a multiparameter group. Suppose that $a_1, a_2, \ldots, a_r$ are the parameters of group elements near the identity element e, and that $(z_1, \ldots z_n)$ are coordinates on M.

Definition 1.6.11 Given a differentiable group action $G \times M \to M$, the *infinitesimals* of the group action are defined to be the derivatives of the $\widetilde{z}_i$

on M with respect to the group parameters a_j evaluated at the identity element e, and are denoted as

$$\frac{\partial \widetilde{z}_i}{\partial a_j}\bigg|_{g=e} = \zeta_j^i. \tag{1.42}$$

Definition 1.6.12 Given a group action of G on $M = X \times U$, the *infinitesimals of the prolonged group action* are defined to be the derivatives of the $\widetilde{u}_K^\alpha$ with respect to the group parameters a_j, evaluated at the identity element e, and are denoted as,

$$\frac{\partial \widetilde{x}_i}{\partial a_j}\bigg|_{g=e} = \xi_j^i, \qquad \frac{\partial \widetilde{u}^\alpha}{\partial a_j}\bigg|_{g=e} = \phi_{,j}^\alpha, \qquad \frac{\partial \widetilde{u}_K^\alpha}{\partial a_j}\bigg|_{g=e} = \phi_{K,j}^\alpha. \tag{1.43}$$

Keeping track of indices rapidly becomes tedious and so we usually compile the infinitesimals in table form, with one row for each group parameter and one column for each coordinate:

	x_i	u^α	u_K^α
a_j	ξ_j^i	$\phi_{,j}^\alpha$	$\phi_{K,j}^\alpha$

Exercise 1.6.13 For the action in Example (1.3.12), show that the table of infinitesimals is

	x	t	u	v	u_K	v_K
α	0	0	$-v$	u	$-v_K$	u_K
a	0	0	1	0	0	0
b	0	0	0	1	0	0

A one parameter subgroup $h(t)$ induces a path in the parameters as $t \mapsto a_j(t)$, $j = 1, \ldots, r$ near the identity element. Then defining α^j, $j = 1, \ldots, r$ by

$$\frac{d}{dt}\bigg|_{t=0} a_j(t) = \alpha^j, \tag{1.44}$$

the chain rule yields

$$\mathbf{v}_h \cdot z_i = \sum_{j=1}^r \zeta_j^i \alpha^j.$$

Moreover for a differentiable function f defined on M we have

$$(\mathbf{v}_h \cdot f)(z_1, \ldots z_n) = \sum_i \frac{\partial f}{\partial z_i} \mathbf{v}_h \cdot z_i = \sum_j \left(\sum_i \zeta_j^i \frac{\partial f}{\partial z_i} \right).$$

Definition 1.6.14 For a differentiable group action $G \times M \to M$, with group parameters $a_1, a_2, \ldots, a_r$ near the identity element e, and $z = (z_1, \ldots, z_n) \in M$, the *infinitesimal vector* corresponding to the group parameter a_j is defined to be

$$\mathbf{v}_j = \sum_{i=1}^{n} \zeta_j^i \frac{\partial}{\partial z_i}, \tag{1.45}$$

and hence

$$\mathbf{v}_h = \sum_j \alpha^j \mathbf{v}_j. \tag{1.46}$$

For readers unfamiliar with differential operators as vectors, see Section 2.2.2.

The infinitesimal action of a *prolonged* one parameter subgroup $h(t)$ is, for α^j given in (1.44),

$$
\begin{aligned}
\mathbf{v}_h \cdot x_i &= \sum_j \xi_j^i \alpha^j \\
\mathbf{v}_h \cdot u^\alpha &= \sum_j \phi_{,j}^\alpha \alpha^j \\
\mathbf{v}_h \cdot u_K^\alpha &= \sum_j \phi_{K,j}^\alpha \alpha^j.
\end{aligned}
\tag{1.47}
$$

The prolonged infinitesimal action on functions is obtained by applying Definition 1.6.7 to functions on the prolonged space,

$$(\mathbf{v}_h \cdot f)(x_i, u^\alpha, u_K^\alpha) = \sum_j \alpha^j \left(\sum_{i,\alpha,K} \xi_j^i \frac{\partial f}{\partial x_i} + \phi_{,j}^\alpha \frac{\partial f}{\partial u^\alpha} + \phi_{K,j}^\alpha \frac{\partial f}{\partial u_K^\alpha} \right) \tag{1.48}$$

while the infinitesimal vector corresponding to the jth group parameter a_j defined in (1.45), for a prolonged action is

$$\mathbf{v}_j = \sum_{i,\alpha,K} \xi_j^i \frac{\partial}{\partial x_i} + \phi_{,j}^\alpha \frac{\partial}{\partial u^\alpha} + \phi_{K,j}^\alpha \frac{\partial}{\partial u_K^\alpha}. \tag{1.49}$$

Exercise 1.6.15 Consider the third of the $SL(2)$ actions in Example 1.2.14,

$$\widetilde{x} = \frac{ax+b}{cx+d}, \qquad \widetilde{y} = 6c(cx+d) + (cx+d)^2 y, \qquad ad - bc = 1.$$

Take local coordinates near the identity to be (a, b, c) so that $e = (1, 0, 0)$. Verify the table of infinitesimals,

	x	y	y_x
a	$2x$	$-2y$	$-4y_x$
b	1	0	0
c	$-x^2$	$6 + 2xy$	$4xy_x + 2y$

Show that this table, together with equation (1.47) for the one parameter subgroup in Exercise 1.5.7, yields the results of Exercise 1.6.6. Hint: $(\alpha, \beta, \gamma) = (\alpha^1, \alpha^2, \alpha^3)$.

Remark 1.6.16 Applying Definition 1.6.10 to the prolonged action is the first step of Sophus Lie's algorithm for calculating the symmetry group of a differential equation. This algorithm is discussed in detail in textbooks, for example Bluman and Cole (1974), Ovsiannikov (1982), Bluman and Kumei (1989), Stephani (1989), Olver (1993), Hydon (2000) and Cantwell (2002), and we refer the interested reader to these.

The infinitesimals and infinitesimal operators defined above are all with respect to given coordinates on M and given parameters describing the group action. Coordinates are the reality of applying the theory and writing software, but it is also important to have the geometric point of view developed in the next two chapters. The full importance of the infinitesimal vectors will become apparent in Chapter 3.

1.6.1 The prolongation formula

Given a prolonged action, it is not necessary to calculate $\widetilde{u}_K^\alpha$ in order to calculate the infinitesimals $\phi_{K,j}^\alpha$.

In the simplest case where we have $u = u(x)$, that is, one dependent and one independent variable, we can obtain the infinitesimal action on y_x without calculating $\widetilde{y}_x$ as follows. Observe that

$$
\begin{aligned}
\frac{\partial \widetilde{u}_x}{\partial g_j} &= \frac{\partial}{\partial g_j} \left(\frac{\mathrm{d}\widetilde{u}}{\mathrm{d}x} \Big/ \frac{\mathrm{d}\widetilde{x}}{\mathrm{d}x} \right) \\
&= \left(\frac{\mathrm{d}\widetilde{x}}{\mathrm{d}x} \right)^{-2} \left(\frac{\mathrm{d}\widetilde{x}}{\mathrm{d}x} \frac{\partial}{\partial g_j} \frac{\mathrm{d}\widetilde{u}}{\mathrm{d}x} - \frac{\mathrm{d}\widetilde{u}}{\mathrm{d}x} \left(\frac{\partial}{\partial g_j} \frac{\mathrm{d}\widetilde{x}}{\mathrm{d}x} \right) \right) \\
&= \left(\frac{\mathrm{d}\widetilde{x}}{\mathrm{d}x} \right)^{-2} \left(\frac{\mathrm{d}\widetilde{x}}{\mathrm{d}x} \frac{\mathrm{d}}{\mathrm{d}x} \frac{\partial \widetilde{u}}{\partial g_j} - \frac{\mathrm{d}\widetilde{u}}{\mathrm{d}x} \frac{\mathrm{d}}{\mathrm{d}x} \frac{\partial \widetilde{x}}{\partial g_j} \right)
\end{aligned}
$$

since derivatives with respect to x and the g_j commute. Evaluating this at $g = e$, when $\mathrm{d}\widetilde{x}/\mathrm{d}x = 1$, yields

$$
\phi_{[x],j} = \frac{\mathrm{d}}{\mathrm{d}x} \phi_{.j} - u_x \frac{\mathrm{d}}{\mathrm{d}x} \xi_j, \tag{1.50}
$$

where we have denoted by $[x]$ the particular index of differentiation on u whose infinitesimal we are considering. Note that the derivative operators in equation (1.50) are *total* derivatives. This is important since typically ξ and ϕ depend on the dependent variables.

Exercise 1.6.17 Adapt the calculation above to show that in the case $u = u(x)$, and $K = [x \ldots x]$, with $|K|$ terms, $Kx = [xx \ldots x]$, with $|K| + 1$ terms,

$$\phi_{Kx,j} = \frac{\mathrm{d}}{\mathrm{d}x}\phi_{K,j} - u_{Kx}\frac{\mathrm{d}}{\mathrm{d}x}\xi_j. \tag{1.51}$$

Exercise 1.6.18 Extend the calculation of the previous exercise to show that if $u = u(x, y)$, $K = [x \ldots xy \ldots y]$, $Kx = [xx \ldots xy \ldots y]$, then

$$\phi_{Kx,j} = \frac{\mathrm{D}}{\mathrm{D}x}\phi_{K,j} - u_{Kx}\frac{\mathrm{D}}{\mathrm{D}x}\xi_j^x - u_{Ky}\frac{\mathrm{D}}{\mathrm{D}x}\xi_j^y \tag{1.52}$$

where

$$\xi_j^x = \frac{\partial}{\partial g_j}\Big|_{g=e}\widetilde{x}, \qquad \xi_j^y = \frac{\partial}{\partial g_j}\Big|_{g=e}\widetilde{y}$$

and

$$\frac{\mathrm{D}}{\mathrm{D}x} = \frac{\partial}{\partial x} + u_x\frac{\partial}{\partial u} + u_{xx}\frac{\partial}{\partial u_x} + u_{xy}\frac{\partial}{\partial u_y} + \cdots = \frac{\partial}{\partial x} + \sum_K u_{Kx}\frac{\partial}{\partial u_K}$$

is the *total* derivative operator in the x direction. Find the matching formula for $\phi_{Ky,j}$.

The formula (1.52) is a recursion formula satisfied by the $\phi_{K,j}$ in the case of two independent and one dependent variables. A more general result follows.

Theorem 1.6.19 *The $\phi_{K,j}^\alpha$ in terms of the ξ_j^i and $\phi_{,j}^\alpha$ is*

$$\phi_{K,j}^\alpha = D^K\left(\phi_{,j}^\alpha - \sum_i u_i^\alpha \xi_j^i\right) + \sum_i \xi_j^i u_{Ki}^\alpha, \tag{1.53}$$

where D^K is the total *derivative of index K, $u_i^\alpha = \partial u^\alpha/\partial x_i$ and $u_{Ki}^\alpha = \partial u_K^\alpha/\partial x_i$.*

Exercise 1.6.20 Verify the table of infinitesimals given below for the action

$$\widetilde{x} = \frac{ax + b}{cx + d}, \qquad \widetilde{u}(\widetilde{x}) = u(x),$$

where $ad - bc = 1$, in two different ways: by calculating $\phi_{K,j}$ directly from $\widetilde{u_K}$, and by using the formulae above

	x	u	u_x	u_{xx}	u_{Kx}
a	$2x$	0	$-2u_x$	$-4u_{xx}$	$-2(|K| + 1)u_{Kx}$
b	1	0	0	0	0
c	$-x^2$	0	$2xu_x$	$4xu_{xx} + 2u_x$	$(|K| + 1)(2u_{Kx} + |K|u_K)$

where $|K|$ is the length of K.

The proof of Theorem 1.6.19 follows from iterative use of the chain rule. These formulae and their derivations can be found in every textbook on symmetries of differential equations, for example Olver (1993) or Bluman and Kumei (1989), in a seemingly endless variety of notational styles. It is well worth taking the time to calculate a selection of prolongations of infinitesimals, not only to be sure which index is which in the preferred notation, but then also to implement it in the preferred computer algebra system. The software will be needed to do the calculations in Chapter 4.

Exercise 1.6.21 Implement the prolongation formulae, equation (1.53). The input will be lists of dependent and independent variables, the infinitesimals ξ_i and ϕ^α and an index of differentiation K. The output will be ϕ_K^α.

Remark 1.6.22 Virtually every computer algebra system has a package that implements Lie's algorithm to find symmetries of differential equations, and all these have, of necessity, implementations of the prolongation formulae buried in them. A review of the software packages available has been given by W. Hereman in Ibragimov (1996), Chapter 13; one recent package is Carminati and Vu (2000).

1.6.2 From infinitesimals to actions

We now state the major theorem of this section, that the one parameter group action can be derived from the infinitesimals, at least for small t.

Theorem 1.6.23 *The solution of the differential system*

$$\frac{\mathrm{d}}{\mathrm{d}t}\widetilde{z}_i = \zeta^i(\widetilde{z}), \qquad i = 1, \ldots, n$$

$$\widetilde{z}\big|_{t=0} = z \tag{1.54}$$

yields a (local) one parameter transformation group whose infinitesimals are ζ^i.

Typically the system can be solved only for t in an interval about 0, in which case a local action results. The proof of this theorem is given in Chapter 2. In the particular case of a prolonged action, the theorem is written as follows.

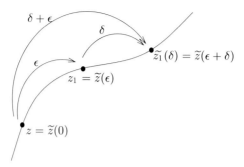

Figure 1.8 The one parameter group condition for a transformation group means: if a point z 'flows' for 'time' ϵ and then for 'time' δ, it arrives at the same point as if it flowed for 'time' $\delta + \epsilon$.

Theorem 1.6.24 *The solution of the differential system*

$$\frac{\mathrm{d}}{\mathrm{d}t}\widetilde{x}_i = \xi^i(\widetilde{x}, \widetilde{u}), \qquad i = 1, \ldots, p$$

$$\frac{\mathrm{d}}{\mathrm{d}t}\widetilde{u}^\alpha = \phi^\alpha(\widetilde{x}, \widetilde{u}), \qquad \alpha = 1, \ldots, q \qquad (1.55)$$

$$(\widetilde{x}, \widetilde{u})\big|_{t=0} = (x, u)$$

yields a one parameter transformation group whose infinitesimals are ξ^i and ϕ^α.

Note that the x_i and the u^α appear on the same footing in the ordinary differential system (1.55). It is only when the induced action on the derivatives is calculated that the u^α are taken to be functions of the x_i.

Remark 1.6.25 In applications where t is one of the existing independent variables, we set the group parameter to be ϵ.

The existence and uniqueness of the solution to first order ordinary differential systems with given initial values is the key result, not only to obtain $(\widetilde{x}, \widetilde{u})$ but to prove the one parameter group property holds. To describe this property for a transformation group, we need some notation. Set $\widetilde{z}(\epsilon) = z_1$. If we now take z_1 as our new initial value, we obtain the curve $\delta \mapsto \widetilde{z_1}(\delta)$. The one parameter group action property is then (see Figure 1.8)

$$z_1 = \widetilde{z}(\epsilon) \implies \widetilde{z_1}(\delta) = \widetilde{z}(\epsilon + \delta).$$

A full discussion is given in Chapter 2.

Typically, it is the infinitesimals that are given in an application, not the group action. Since it is hard to integrate systems like (1.55), we will be trying to avoid this integration step as much as possible. One of the great features of the moving frame theory is that so many of the calculations can be done with only the infinitesimals. In addition, it is important to ask yourself, for the particular application at hand, 'if the integration step can only be handled approximately, does an approximate solution suffice?'

The uniqueness of the solution of the differential system implies that if you start with a one parameter group action, obtain the infinitesimals, and then integrate, you obtain the same one parameter group you started with. If you do not start with an action satisfying the one parameter group property, then the solution of the system is a reparametrisation of the group action that does satisfy it.

Exercise 1.6.26 Consider the scaling transformation $\tilde{x} = \lambda^2 x$ which is an action of $(\mathbb{R}^+, \cdot)$, whose identity element is $\lambda = 1$. Differentiating $\tilde{x}$ with respect to λ at $\lambda = 1$ yields the infinitesimal $\xi = 2x$.
What to do: integrating the system $d\tilde{x}/dt = 2\tilde{x}$ with initial condition $\tilde{x} = x$ at $t = 0$ yields $\tilde{x} = \exp(2t)x$. Show this is a reparametrisation of the scaling transformation that satisfies the one parameter group property.
What not to do: integrating the system $d\tilde{x}/d\lambda = 2\tilde{x}$ with initial condition $\tilde{x} = x$ at $\lambda = 1$ yields $\tilde{x} = \exp(2(\lambda - 1))x$. Show this is not a group action of $(\mathbb{R}^+, \cdot)$.

The purpose of the next exercise is to give the reader a taste of the skullduggery required to solve a typical differential system and verify Theorem 1.6.24 in practice. These infinitesimals arose in a study of non-classical reductions of the equation $u_t = u_{xx} + f(u)$, for $f(u)$ a cubic (Clarkson and Mansfield, 1993).

Exercise 1.6.27 Find the group actions corresponding to the given infinitesimals,

	x	t	u
ϵ	$3\mu \tan(\mu x + \kappa)$	1	$-\mu^2(3u + b)\sec^2(\mu x + \kappa)$

where μ, κ and b are constants. (Since t is here an independent variable, we use ϵ for the group parameter.) Verify the group action property, and show that $\mathcal{I} = \exp(-3\mu^2 t)\sin(\mu x + \kappa)$ and $\mathcal{J} = (3u + b)\tan(\mu x + \kappa)$ are invariants.

Solution to Exercise 1.6.27   We need to solve

$$\widetilde{x}_\epsilon = 3\mu \tan(\mu\widetilde{x} + \kappa)$$
$$\widetilde{t}_\epsilon = 1$$
$$\widetilde{u}_\epsilon = -\mu^2(3\widetilde{u} + b)\sec^2(\mu\widetilde{x} + \kappa)$$

together with $\widetilde{x}(0) = x$, $\widetilde{t}(0) = t$, $\widetilde{u}(0) = u$. The first two equations are easily integrated to give

$$\sin(\mu\widetilde{x} + \kappa) = \exp(3\mu^2\epsilon)\sin(\mu x + \kappa), \qquad \widetilde{t} = t + \epsilon.$$

Back-substituting $\epsilon = \widetilde{t} - t$ into the first expression and rearranging terms, we obtain that $\exp(-3\mu^2 t)\sin(\mu x + \kappa) = \exp(-3\mu^2\widetilde{t})\sin(\mu\widetilde{x} + \kappa)$, in other words, $\mathcal{I}$ is an invariant. To verify the group action property for the variable x, set $x_1 = \widetilde{x}(\epsilon)$. Note that $\sin(\mu x_1 + \kappa) = \exp(3\mu^2\epsilon)\sin(\mu x + \kappa)$ and $\sin(\mu\widetilde{x_1}(\delta) + \kappa) = \exp(3\mu^2\delta)\sin(\mu x_1 + \kappa)$ and thus

$$\begin{aligned}
\sin(\mu\widetilde{x_1}(\delta) + \kappa) &= \exp(3\mu^2\delta)\sin(\mu x_1 + \kappa) \\
&= \exp(3\mu^2\delta)\exp(3\mu^2\epsilon)\sin(\mu x + \kappa) \\
&= \exp(3\mu^2(\epsilon + \delta))\sin(\mu x + \kappa) \\
&= \sin(\mu\widetilde{x}(\epsilon + \delta) + \kappa)
\end{aligned}$$

so that $\widetilde{x_1}(\delta) = \widetilde{x}(\epsilon + \delta)$ as required (for small enough δ and ϵ). Verifying the flow condition for the variable t is trivial and is left to the reader. To simplify the integration of the third equation, note that $\widetilde{x}_{\epsilon\epsilon} = 3\mu^2\sec^2(\mu\widetilde{x} + \kappa)\widetilde{x}_\epsilon$ and thus the third equation becomes

$$\frac{3\widetilde{u}_\epsilon}{3\widetilde{u} + b} = -\frac{\widetilde{x}_{\epsilon\epsilon}}{\widetilde{x}_\epsilon}$$

which is easily integrated to give

$$c = \widetilde{x}_\epsilon(3\widetilde{u} + b) = 3\mu \tan(\mu\widetilde{x} + \kappa)(3\widetilde{u} + b)$$

where c is a constant of integration. Thus the quantity $\mathcal{J} = \tan(\mu\widetilde{x} + \kappa)(3\widetilde{u} + b)$ is an invariant; this necessarily has the same value for all values of ϵ, so we obtain

$$\tan(\mu\widetilde{x} + \kappa)(3\widetilde{u} + b) = \tan(\mu x + \kappa)(3u + b).$$

Solving this for $\widetilde{u}$ and back-substituting for $\widetilde{x}$ in terms of ϵ and x yields the desired expression for $\widetilde{u}(\epsilon)$ in terms of ϵ and the initial position, (x, t, u). Verifying the group property using this expression is, however, not recommended.

Instead, we use the invariant $\mathcal{J}$. We have

$$\tan(\mu\widetilde{x_1}(\delta) + \kappa)(3\widetilde{u_1}(\delta) + b) = \tan(\mu x_1 + \kappa)(3u_1 + b)$$

$$\text{by invariance of } \mathcal{J}$$

$$= \tan(\mu\widetilde{x}(\epsilon) + \kappa)(3\widetilde{u}(\epsilon) + b)$$

$$\text{as } x_1 = \widetilde{x}(\epsilon), u_1 = \widetilde{u}(\epsilon)$$

$$= \tan(\mu\widetilde{x}(\epsilon + \delta) + \kappa)(3\widetilde{u}(\epsilon + \delta) + b)$$

$$\text{by invariance of } \mathcal{J}$$

$$= \tan(\mu\widetilde{x_1}(\delta) + \kappa)(3\widetilde{u}(\epsilon + \delta) + b)$$

using the group property for the variable x in the last step. Thus $\widetilde{u_1}(\delta) = \widetilde{u}(\epsilon + \delta)$ as required. The invariance of $\mathcal{J}$ can also be checked directly by noting that $d\mathcal{J}/d\epsilon \equiv 0$, and similarly for the other invariant. We leave this to the reader. $\qquad\square$

2
Calculus on Lie groups

In this chapter we examine briefly the details of the technical definition of a Lie group. This chapter can be skipped on a first reading of this book. Eventually, however, taking a small amount of time to be familiar with the the concepts involved will pay major dividends when it comes to understanding the proofs of the key theorems.

By definition, Lie groups are locally Euclidean, so we can use tools we know and love from calculus to study functions, vector fields and so on that can be defined on them. Thus, we study differentiation on a Lie group. There are at least three important cases to consider. The first involves understanding the *intrinsic* definition of tangent vectors. These ideas inform every other understanding of a tangent vector, so we do that first. A second and simpler line of argument is strictly for matrix presentations, while a third treats tangent vectors as linear, first order differential operators. We will need all three.

The major theorem we prove is that the set of tangent vectors at any given point $g \in G$ is in one-to-one correspondence with the set of one parameter subgroups of G. After a discussion of the exponential map in its various guises, we end the chapter with a discussion of concepts analogous to tangent vectors, one parameter subgroups and the exponential map for transformation groups.

2.1 Local coordinates

The technical definition of a Lie group is that it is a smooth, locally Euclidean space, such that the multiplication and inverse maps are smooth. In this section, we describe what this all means.

Up close, every small enough piece of an n dimensional Lie group looks like a piece of $\mathbb{R}^n$ in the sense that one has coordinates, as in Figure 2.1. Thus we say that Lie groups are *locally Euclidean*.

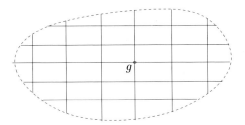

Figure 2.1 Near every $g \in G$, one can define a system of coordinates centred at g.

Definition 2.1.1 A *chart* or *coordinate map* for the Lie group G, is given by an invertible map

$$\phi : \mathcal{U}(g) \longrightarrow \phi(\mathcal{U}(g)) \subset \mathbb{R}^n,$$

where $\mathcal{U}(g) \subset G$ is a neighbourhood of g and $\phi(\mathcal{U}(g))$ is an open subset of $\mathbb{R}^n$ for some n. If in addition, $\phi(g) = 0$, we say the chart is *centred at* $g \in G$.

Remark 2.1.2 A subset U of $\mathbb{R}^n$ is *open* if for every $x \in U$, there is an $\epsilon > 0$ such that the ϵ-ball, $B_\epsilon(x) = \{y \in \mathbb{R}^n \mid |x - y| < \epsilon\}$ is contained wholly within U. This requirement relieves us from considering singularities of coordinate maps which occur at the boundaries of their domains of definition, as well as ensuring that the image of the coordinate map is fully n dimensional.

The best known example of a locally Euclidean space which is not globally Euclidean is the the surface of a sphere; think of the globe. (What are the coordinate maps for the surface of the earth?) Unsurprisingly, the coordinate maps are known as *charts* and the set of all possible charts is called an *atlas*.

In addition to coordinate charts existing about every point $g \in G$, we also need the charts to be smoothly coherent in the following sense.

Definition 2.1.3 We shall call a set of charts $\{(\mathcal{U}_j, \phi_j) \mid j \in \mathcal{J}\}$ for an n dimensional space M a smooth *atlas* if

(i)

$$\bigcup_{i \in \mathcal{J}} \mathcal{U}_i = M$$

and

(ii) for all $i, j \in \mathcal{A}$ such that $\mathrm{domain}(\phi_i) \cap \mathrm{domain}(\phi_j) \neq \emptyset$, then the *interchange map*,

$$\phi_j \circ \phi_i^{-i} : \phi_i(\mathrm{domain}(\phi_i) \cap \mathrm{domain}(\phi_j)) \subset \mathbb{R}^n \longrightarrow \mathbb{R}^n,$$

is smooth, that is, infinitely differentiable, according to the standard definitions given in several variable calculus.

One consequence of the definition is that the dimensions of the image space of the coordinate maps are all the same.

Definition 2.1.4 A space with a smooth atlas is called a smooth *manifold*.

The second part of the definition of a Lie group is that multiplication maps and the inverse map are smooth.

Definition 2.1.5 Given $h \in G$, left multiplication by h is defined by

$$L_h : G \longrightarrow G, \qquad g \mapsto hg,$$

while right multiplication by h is defined by

$$R_h : G \longrightarrow G, \qquad g \mapsto gh.$$

Definition 2.1.6 We say left multiplication is a smooth map provided for every $h \in G$ and every pair of coordinate maps ϕ_i, ϕ_j satisfying

$$L_h \text{domain}(\phi_i) \cap \text{domain}(\phi_j) \neq \emptyset,$$

that the map $\phi_j \circ L_h \circ \phi_i^{-1}$ is smooth, according to the standard definitions of smoothness in several variable calculus, on its domain of definition.

The definitions for right multiplication and the inverse map to be smooth are similar. Since multiplication is smooth, it follows that given coordinates near the identity, we obtain coordinates near *any* element of the Lie group using left or right multiplication. If $\mathcal{U}(e)$ is a neighbourhood of $e \in G$, then $\mathcal{U}(g) = g \cdot \mathcal{U}(e)$ is a neighbourhood of g. Moreover, if $\phi : \mathcal{U}(e) \longrightarrow \mathbb{R}^n$ is a coordinate map, then $\phi \circ g^{-1}$ is a coordinate map for $\mathcal{U}(g)$. One can also use right multiplication; this will yield a different chart in general.

Definition 2.1.7 A *Lie group* G is a smooth manifold which is also a group, such that the multiplication and inverse maps are smooth.

Example 2.1.8 The unit circle, $S^1 \subset \mathbb{C}$ is a Lie group with multiplication,

$$\exp(i\theta) \exp(i\psi) = \exp(i(\theta + \psi)).$$

For $\exp(i\theta) \in S^1$, and $0 < \epsilon < \pi$, take

$$\mathcal{U}(\exp(i\theta)) = \{\exp(i\psi) \mid |\theta - \psi| \leq \epsilon\}$$

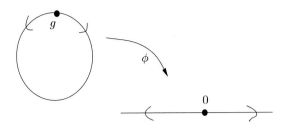

Figure 2.2 $\phi : \mathcal{U}(g) \longrightarrow \phi(\mathcal{U}(g)) \subset \mathbb{R}, \phi(g) = 0.$

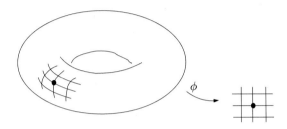

Figure 2.3 $\phi : \mathcal{U}(g) \longrightarrow \phi(\mathcal{U}(g)) \subset \mathbb{R}^2, \phi(g) = 0.$

and then the coordinate map ϕ centred at θ is given by (see Figure 2.2)

$$\phi(\exp(\mathrm{i}\psi)) = \psi - \theta.$$

Exercise 2.1.9 Consider the coordinate maps given in Example 2.1.8 above. For coordinate maps ϕ_0 and ϕ_1 centred at θ_0 and θ_1 in $S^1 \subset \mathbb{C}$ respectively, whose domains have non-trivial intersection, show that the interchange map is a translation,

$$\phi_1 \circ \phi_0^{-1}(\psi) = \psi + \theta_0 - \theta_1.$$

Example 2.1.10 The two dimensional torus, $S^1 \times S^1$, is a Lie group, with the product multiplication,

$$(\exp(\mathrm{i}\theta), \exp(\mathrm{i}\varphi)) \cdot (\exp(\mathrm{i}\chi), \exp(\mathrm{i}\psi)) = (\exp(\mathrm{i}(\theta + \chi)), \exp(\mathrm{i}(\varphi + \psi))).$$

Exercise 2.1.11 Define a neighbourhood and a coordinate chart for any element of $S^1 \times S^1$, see Figure 2.3.

Example 2.1.12 Elements of the Lie group $SL(2, \mathbb{R})$ near the identity are given by

$$\begin{pmatrix} a & b \\ c & \dfrac{1 + bc}{a} \end{pmatrix}$$

since a is close to unity and is therefore non-zero. Coordinates in this neighbourhood of the identity I_2, centred at I_2, are then (for example)

$$\phi : \begin{pmatrix} a & b \\ c & \dfrac{1+bc}{a} \end{pmatrix} \mapsto (a-1, b, c) \in \mathbb{R}^3.$$

Exercise 2.1.13 Write down coordinates for a neighbourhood of *any* element of $SL(2, \mathbb{R})$ by first using left (or right) multiplication to take the given element to the identity, followed by a coordinate map at the identity. Given two such charts, what is the interchange map?

Exercise 2.1.14 Write down coordinates for elements of $SO(3)$ near the identity, using both Euler angles and the Cayley parametrisation (Fässler and Stiefel, 1992, page 83).

2.2 Tangent vectors on Lie groups

Using the system of coordinates we can talk about tangent vectors on G. Given the wealth of ways Lie groups present themselves, the naive view of a tangent vector as an arrow sitting on a surface is not particularly helpful, unless you happen to have a surface such as a two dimensional torus. Instead, the best way to think of a tangent vector is in terms of *paths*; *to obtain a tangent vector, take a smooth path and differentiate it.*

In standard Differential Topology, one differentiates the image of the path under a coordinate map, since differentiation of a path in $\mathbb{R}^n$ is well defined. But it is enough to think of the path itself as representing the vector; this is the so-called *intrinsic* definition of a vector. For Lie groups in their matrix representation, one can describe tangent vectors using matrices, taking advantage of the fact that the set of $n \times n$ matrices is a linear space. Finally, we look at tangent vectors as first order differential operators. These last two descriptions are particularly useful in applications.

Definition 2.2.1 A *smooth path with basepoint* $g \in G$ is a map $\gamma : [-\epsilon, \epsilon] \subset \mathbb{R} \longrightarrow G$ for some $\epsilon > 0$, such that $\gamma(0) = g$, and wherever γ maps into the domain of a coordinate map ϕ, then $\phi \circ \gamma$ is a smooth curve in $\mathbb{R}^n$ (Figure 2.4).

While we have defined a path as a map, we also speak of its image $\gamma([-\epsilon, \epsilon]) \subset G$ as being the path, depending on the context.

If $\gamma : t \mapsto \gamma(t) \subset \mathcal{U}(g) \subset G$ is a path with basepoint $g \in G$, and ϕ is a coordinate map centred at g, then $\phi(\gamma(t))$ is a path with basepoint $0 \in \mathbb{R}^n$.

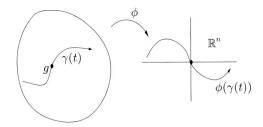

Figure 2.4   A coordinate map centred at g takes each path through g to a path through 0.

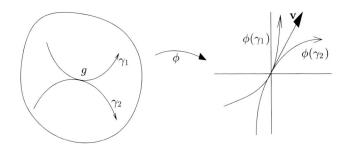

Figure 2.5   Paths are equivalent at g if their tangent vectors at $\phi(g)$ are equal.

The path $\phi \circ \gamma$ has a tangent vector at $0 \in \mathbb{R}^n$ given by $\mathbf{v} = (\phi \circ \gamma)'(0)$. The idea is to think of a vector at g as a curve based at g, and to differentiate it in coordinates to produce a traditional vector when necessary. Unfortunately, more than one path at g leads to the same $\mathbf{v}$ in $\mathbb{R}^n$. In order to have a technically rigorous definition of a tangent space, we need to view as equivalent all paths giving the same vector in $\mathbb{R}^n$ under the same coordinate map. The equivalence relation is pictured in Figure 2.5.

Definition 2.2.2   Two paths γ_1, $\gamma_2 \subset \mathcal{U}(g)$ both with basepoint g, that is, satisfying $g = \gamma_1(0) = \gamma_2(0)$, are said to be *equivalent* at g, $\gamma_1 \sim \gamma_2$, if for some coordinate map ϕ with $g \in \mathrm{domain}(\phi)$,

$$\left. \frac{\mathrm{d}}{\mathrm{d}t} \right|_{t=0} \phi \circ \gamma_1 \;=\; \left. \frac{\mathrm{d}}{\mathrm{d}t} \right|_{t=0} \phi \circ \gamma_2.$$

Note that part of the definition of equivalence is that their basepoints, $\gamma_i(0)$ are equal. The equivalence class of γ is denoted $[\gamma]$.

Paths that are equivalent in one coordinate chart are equivalent in every coordinate chart; since $\phi_1 \circ \gamma = (\phi_1 \circ \phi^{-1}) \circ \phi \circ \gamma$, the result follows by applying the chain rule.

Definition 2.2.3 The *tangent space* at $g \in G$, denoted $T_g G$, is the set of all equivalence classes of smooth paths γ with basepoint $g = \gamma(0)$.

Theorem 2.2.4 *The tangent space $T_g G$ is a linear space.*

Indeed, let ϕ be a coordinate map centred at g, that is, $\phi(g) = 0$. If γ_1, γ_2 are paths with basepoint g then define $\gamma_1 + \gamma_2$ to be the path

$$\phi^{-1}(\phi \circ \gamma_1 + \phi \circ \gamma_2)$$

and for $k \in \mathbb{R}$, define $k\gamma$ to be the path

$$\phi^{-1}(k\phi \circ \gamma).$$

Note that $\gamma_1 + \gamma_2$ and $k\gamma$ have their basepoints at g.

Exercise 2.2.5 Check that if $\gamma_1 \sim \gamma_1'$ and $\gamma_2 \sim \gamma_2'$ then $(\gamma_1 + \gamma_2) \sim (\gamma_1' + \gamma_2')$. Similarly, if $\gamma \sim \gamma'$ then $(k\gamma) \sim (k\gamma')$. Conclude that addition and multiplication by a scalar are well defined on equivalence classes. Hence prove Theorem 2.2.4.

If M and N are manifolds and $f : M \to N$ is a map, then a smooth path γ on M maps to a path $f \circ \gamma$ on N.

Definition 2.2.6 We say f is differentiable at $x \in M$ if the *tangent map*

$$T_x f : T_x M \to T_{f(x)} N, \qquad [\gamma] \mapsto [f \circ \gamma]$$

is well defined (Figure 3.6). If $T_x f$ exists for every $x \in M$, we say f is differentiable on M and write the tangent map as $Tf : TM \to TN$, where $Tf(v) = T_x(v)$ for $v \in T_x M$.

In coordinates, f is differentiable at x if for every differentiable path γ based at $x \in M$, the path $f \circ \gamma$ is differentiable on N. The tangent map is then

$$\frac{\mathrm{d}}{\mathrm{d}t}\bigg|_{t=0} \gamma(t) \mapsto \frac{\mathrm{d}}{\mathrm{d}t}\bigg|_{t=0} f(\gamma(t)).$$

Recall that for G to be a Lie group, left multiplication (Definition 2.1.5) must be a smooth map.

Exercise 2.2.7 Show that

$$\gamma \sim \gamma' \implies h \cdot \gamma \sim h \cdot \gamma'.$$

Conclude that left multiplication induces a map

$$T_g L_h : T_g G \longrightarrow T_{hg} G, \qquad [\gamma] \mapsto [h \cdot \gamma].$$

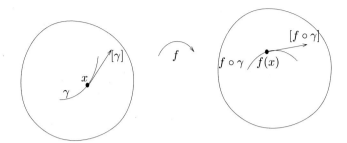

Figure 2.6 The definition of the tangent map $T_x f$, which sends the vector $[\gamma]$ at x to the vector $T_x f([\gamma]) = [f \circ \gamma]$ at $f(x)$.

If h is sufficiently close to the identity so that $hg \in \mathcal{U}(g)$, then we can use the same coordinates to examine both γ and $h \cdot \gamma$. Then $T_g L_h$ (defined in Exercise 2.2.7) is represented by the Jacobian of $\phi L_h \phi^{-1}$:

$$\frac{\mathrm{d}}{\mathrm{d}t}\bigg|_{t=0} \phi \circ L_h \circ \gamma(t) = D(\phi \circ L_h \circ \phi^{-1}) \frac{\mathrm{d}}{\mathrm{d}t}\bigg|_{t=0} (\phi \circ \gamma)(t).$$

Similar remarks apply to right multiplication and the inverse map.

2.2.1 Tangent vectors for matrix Lie groups

If the Lie group is represented by matrices, then since the space $M_n(\mathbb{R})$ $(M_n(\mathbb{C}))$ of $n \times n$ matrices with real (complex) coefficients is linear, we can differentiate a path in M_n without needing a coordinate map:

$$\frac{\mathrm{d}}{\mathrm{d}t}\bigg|_{t=0} (a_{ij}(t)) = (a'_{ij}(0)).$$

Thus, for example, differentiating paths in $SO(2)$, we obtain

$$\frac{\mathrm{d}}{\mathrm{d}t}\bigg|_{t=0} \begin{pmatrix} \cos \theta(t) & -\sin \theta(t) \\ \sin \theta(t) & \cos \theta(t) \end{pmatrix} = \begin{pmatrix} -\sin \theta(0) & -\cos \theta(0) \\ \cos \theta(0) & -\sin \theta(0) \end{pmatrix} \theta'(0)$$

where multiplication of a matrix by a scalar is understood to be

$$c(a_{ij}) = (ca_{ij}).$$

For a matrix group, the tangent map to left multiplication $T L_A$ is still actually left multiplication by A, since

$$\frac{\mathrm{d}}{\mathrm{d}t}\bigg|_{t=0} A B(t) = A \frac{\mathrm{d}}{\mathrm{d}t}\bigg|_{t=0} B(t)$$

for every path $B(t)$ in the group.

Exercise 2.2.8 Show that the tangent space at the identity of the Lie group $SO(2)$ is

$$T_e(SO(2)) = \left\{ \begin{pmatrix} 0 & -c \\ c & 0 \end{pmatrix} \mid c \in \mathbb{R} \right\}.$$

Hint: let $c = \theta'(0)$. Show that the tangent space at an arbitrary element of $A \in SO(2)$ is given by

$$T_A(SO(2)) = \left\{ A \cdot \begin{pmatrix} 0 & -c \\ c & 0 \end{pmatrix} \mid c \in \mathbb{R} \right\} = \left\{ \begin{pmatrix} 0 & -d \\ d & 0 \end{pmatrix} \cdot A \mid d \in \mathbb{R} \right\}.$$

Show that $T_A(SO(2))$ is a linear space.

Exercise 2.2.9 From $A(t)A(t)^{-1} = I$, find a formula for $(A^{-1})'(t)$ in terms of $A'(t)$. Note that taking the inverse and differentiation are not commutative operations. If $A(0) = I$, show $(A^{-1})'(0) = -A'(0)$. Hint: the product rule is valid for matrices.

Exercise 2.2.10 Prove that the tangent space at the identity of $SU(2)$ consists of matrices of the form

$$\begin{pmatrix} a & b \\ -\overline{b} & -a \end{pmatrix},$$

where $\mathrm{Re}(a) = 0$. Hint: recall from Example 1.1.11 that $A \in SU(2)$ means that

$$A = \begin{pmatrix} \alpha & \beta \\ -\overline{\beta} & \overline{\alpha} \end{pmatrix}$$

with $\alpha\overline{\alpha} + \beta\overline{\beta} = 1$. Differentiate $\alpha(t)\overline{\alpha(t)} + \beta(t)\overline{\beta(t)} = 1$ at $t = 0$, and set $\alpha(0) = 1$, $\beta(0) = 0$, $\alpha'(0) = a$ and $\beta'(0) = b$.

Exercise 2.2.11 The unitary group $U(n)$ is the set of $n \times n$ invertible complex matrices given by

$$U(n) = \{U \in GL(n, \mathbb{C}) \mid \bar{U}^T U = I_n\}.$$

Prove that the tangent space at the identity of $U(n)$, which is denoted $\mathfrak{u}(n)$, consists of matrices which satisfy the equation $A^T + \overline{A} = 0$. Show this set of matrices is a vector space. What is its dimension? Hint: differentiate $\overline{U}(t)^T U(t) = I_n$ with respect to t and set $A = U'(0)$.

2.2.2 Some standard notations for vectors and tangent maps in coordinates

Vectors understood as paths is the all purpose notion from which all the others derive. In $\mathbb{R}^n$, that is, in coordinates, one can calculate the derivative of the path to obtain the standard notion of a vector as an element of $\mathbb{R}^n$ together with a base point. Usually the basepoint is implicit.

If we have a vector $V(z)$ at every point $z \in U \subset \mathbb{R}^n$, then the basepoint for the vector $V(z)$ is z, and we say the map $z \mapsto V(z)$ is a *vector field*.

There are three standard notations for vectors in coordinates. The first is as a *row vector*, used for graphical purposes.

Example 2.2.12 Plot the vector field on $\mathbb{R}^2$, given by $V(x, y) = (-y, x)$ and show it is the infinitesimal action of the usual action of $SO(2)$ in the plane.

A second notation is as a *column vector*, used when doing multivariable calculus. For example, the derivative of a function $f : \mathbb{R}^2 \to \mathbb{R}^2$ in the direction of the vector $V(x, y) = (-y, x)^T$ is

$$Df(V) = \begin{pmatrix} f_x^1 & f_y^1 \\ f_x^2 & f_y^2 \end{pmatrix} \begin{pmatrix} -y \\ x \end{pmatrix}. \tag{2.1}$$

Exercise 2.2.13 To see how (2.1) relates to the definition of a tangent map, let $\gamma(t) = (\cos(t)x - \sin(t)y, \sin(t)x + \cos(t)y)$, so that $(d/dt)|_{t=0}\, \gamma(t) = V$. Compare (2.1) with $(d/dt)|_{t=0}\, f(\gamma(t))$ expressed as a column vector.

The third notation is as a *differential operator*. Thus $V(x, y) = (-y, x)$ is represented by

$$V(x, y) \equiv -y\partial_x + x\partial_y.$$

More generally,

$$(V_1(z), \ldots, V_n(z)) \equiv V_1\partial_{z_1} + \cdots + V_n\partial_{z_n}.$$

We then have that the derivative of a function $f : M \to \mathbb{R}$ in the direction V is $V(f)$.

Exercise 2.2.14 Show $V(f) = Df(V)$, where on the left, V is in operator notation, while on the right, V is a column vector.

Remark 2.2.15 Some authors take the operator notation for a vector as the starting point for the definition of a vector as a *derivation* on the set of smooth functions $C^\infty(M, \mathbb{R})$ to itself (where 'smooth' means, in effect, that a derivation can act). A derivation V is a linear map satisfying $V(fg) = V(f)g + fV(g)$. This definition lacks the intuition and the computational 'nous' inherent in the 'vectors as curves' view, and hides from view the multivariable calculus needed to compute anything.

There are times, however, when the operator notation has a computational advantage. If a function F is an invariant of a group action and V is the infinitesimal action, then with V in operator notation, we have $V(F) = 0$. In other words, we can write the condition for F to be invariant as a differential equation. Further, the operator notation will prove extremely useful in calculating the Lie bracket of two vector fields discussed in the next chapter.

We next consider the tangent map in coordinates. We already saw an example in Exercise 2.2.13 which we now generalise. Consider $f : M \to N$ in some coordinate chart, and suppose V is expressed as a column vector. Then the tangent map $T_x f$ is given by the Jacobian of f evaluated at x.

Definition 2.2.16 If the map f is given in local coordinates as

$$x = (x_1, x_2, \ldots, x_n) \mapsto (f^1(x), f^2(x), \ldots, f^m(x))$$

then the Jacobian of f evaluated at x is the matrix

$$Df(x) = \begin{pmatrix} f^1_{x_1} & f^1_{x_2} & \cdots & f^1_{x_n} \\ f^2_{x_1} & f^2_{x_2} & \cdots & f^2_{x_n} \\ \vdots & \vdots & \ddots & \vdots \\ f^m_{x_1} & f^m_{x_2} & \cdots & f^m_{x_n} \end{pmatrix}.$$

Indeed, if $V = \gamma'(0) = (\gamma_1'(0), \ldots \gamma_n'(0))^T$, then

$$\frac{d}{dt}\Big|_{t=0} f(\gamma(t)) = Df\gamma'(0) = \begin{pmatrix} f^1_{x_1} & f^1_{x_2} & \cdots & f^1_{x_n} \\ f^2_{x_1} & f^2_{x_2} & \cdots & f^2_{x_n} \\ \vdots & \vdots & \ddots & \vdots \\ f^m_{x_1} & f^m_{x_2} & \cdots & f^m_{x_n} \end{pmatrix} \begin{pmatrix} \gamma_1'(0) \\ \gamma_2'(0) \\ \vdots \\ \gamma_n'(0) \end{pmatrix} \qquad (2.2)$$

by the chain rule.

Calculating tangent maps is a little trickier in operator notation. We use the fact that if $\chi : N \to \mathbb{R}$ is a scalar function, then $V(\chi) = D\chi(V)$ (Exercise 2.2.14) and thus

$$(Df(V))(\chi) = D\chi(Df(V)) = D(\chi \circ f)(V) = V(\chi \circ f).$$

For example, if $V = v_1 \partial_x + v_2 \partial_y$ is a vector on $\mathbb{R}^2$ and $f : \mathbb{R}^2 \to \mathbb{R}^2$, then by the chain rule,

$$\begin{aligned} (Df(V))(\chi) &= V(f \circ \chi) \\ &= ((v_1 f^1_x + v_2 f^1_y)\partial_{f^1} + (v_1 f^2_x + v_2 f^2_y)\partial_{f^2})\chi \qquad (2.3) \\ &= (V(f^1)\partial_{f^1} + V(f^2)\partial_{f^2})\chi. \end{aligned}$$

It can be seen that the change of basepoint is encoded in the indices of the new operators.

Exercise 2.2.17 Let $X = \partial_x$ and $Y = x\partial_x$ be vectors at $x \in \mathbb{R}$. Show that if $f : \mathbb{R} \to \mathbb{R}$ is the exponential map, $f(x) = y = \exp(x)$, then $Df(X) = y\partial_y$ and $Df(Y) = y \log(y)\partial_y$.

2.3 Vector fields and integral curves

To prove the main theorems of this chapter and the next, we will need the notion of an integral of a vector field.

The idea of a vector field is familiar to anyone who has seen a weather map depicting, at each point in some region of the earth's surface, the direction and magnitude of the wind. Both the direction and magnitude vary smoothly over the region, and it is not hard to imagine that a speck of dust, dropped into the atmosphere at some point, will be carried along by the wind and thus trace out a path parametrised by time t. This path is called the *flowline* or *integral curve* of the vector field passing through the *initial point*, where the speck was released. The set of such paths represents the intuitive notion of the integral of the vector field. In this section we consider vector fields whose components do not depend explicitly on time. Such vector fields are called *autonomous*.

Remark 2.3.1 Non-autonomous vector fields can be converted to autonomous ones by extending the base space M to $M \times \mathbb{R}$, setting t to be the new coordinate, and extending vectors V to $(V, 1)$ (row notation) or $V + \partial_t$ (operator notation). Thus the theorems we describe here can be extended to the non-autonomous case.

Vector fields are maps from M to the tangent bundle of M, which we now define.

Definition 2.3.2 The *tangent space* of a manifold M is

$$TM = \bigcup_{z \in M} T_z M.$$

Coordinate charts $(\mathcal{U}, \phi)$ on M extend to coordinate charts on TM, with $[\gamma] \in T_z M$ mapping to $(\phi(z), (d/dt)|_{t=0}\, \phi(\gamma(t))) \in \mathbb{R}^n \times \mathbb{R}^n$ where $n = \dim(M)$.

Remark 2.3.3 It is not true in general that the tangent space to a manifold M is the product $M \times \mathbb{R}^n$. The standard counterexample is the two dimensional sphere; if TS^2 were equal to $S^2 \times \mathbb{R}^2$, then there would exist a vector field on S^2 which is everywhere non-zero; this is famously false by the 'hairy ball theorem' (Munkres, 1975, Theorem 10.4).

Definition 2.3.4 A *vector field* on the manifold M is a map

$$V : M \to TM, \qquad V(z) \in T_z M.$$

On the space M itself, at any given point, a vector is an equivalence class of curves, so a vector field gives a class of curves at every point. Under a coordinate map, a vector field can be represented in the traditional way by an arrow at every point. Vectors are also denoted as **v** or X.

Example 2.3.5 Show that there exist everywhere non-zero vector fields on Lie groups. Hint: show there is an everywhere non-zero vector field by taking a vector V at the identity and considering the map $g \mapsto T_e R_g(V)$. This implies that for Lie groups, $TG = G \times \mathbb{R}^r$ where $r = \dim(G)$ (Hirsch, 1976, Exercise 4, page 92).

Definition 2.3.6 A *flowline* or *integral curve* of a vector field V on M is a path $\Gamma : s \mapsto M$ such that for every $s \in \mathbb{R}$, $V(\Gamma(s)) \sim \Gamma(s)$ (recall Definition 2.2.2). A flowline for V satisfying $\Gamma(0) = z$ is denoted by either $s \mapsto \Gamma_s^V(z)$ or if V is understood, by $s \mapsto \Gamma(z, s)$.

Remark 2.3.7 It can happen that integral curves are not defined for all $s \in \mathbb{R}$, in which case the vector field is said to generate only a *partial flow*. As an example, consider the vector field on $\mathbb{R}$ given by $V(x) = x^2 \partial_x$, for which $\Gamma_s^V(x) = x/(1 - sx)$. Thus one can prove only that a flow exists for s in some open interval about $0 \in \mathbb{R}$.

Integral curves for a vector field, with intrinsically defined vectors, are depicted in Figure 2.7. A coordinate map ϕ, with domain $\mathcal{U}$, maps V to a vector field $\overline{V}$ on $\phi(\mathcal{U}) \subset \mathbb{R}^n$, depicted in Figure 2.8, by taking

$$\overline{V}(\phi(z)) = \left.\frac{\mathrm{d}}{\mathrm{d}t}\right|_{t=0} \phi(\gamma(t)), \qquad V(z) = [\gamma].$$

A vector field is continuous if its components in any coordinate system are continuous functions. We state the theorem guaranteeing existence and uniqueness of integral curves of continuous vector fields in its usual form, that is, in coordinates. The integral curves on M are obtained by pulling back the flowlines guaranteed by the theorem for the vector field in $\mathbb{R}^n$, via the coordinate maps. Uniqueness of the flowlines means that integral curves in different coordinate patches match up, so that they can be continued over all of M.

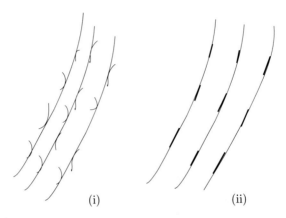

Figure 2.7 Flowlines for an intrinsic vector field on M are everywhere tangent to representative paths at their base points. In (i) arbitrary representative paths of vectors are shown. In (ii), the flowlines themselves represent the vectors at every point (some shown in bold for illustrative purposes).

Figure 2.8 Flowlines for a vector field in $\mathbb{R}^2$.

Theorem 2.3.8 (EUSODE) *Given a continuous vector field V in an open neighbourhood $U \subset \mathbb{R}^n$, then for each $x \in U$, there exists a unique integral curve $s \mapsto \Gamma(x, s)$, for s in some open interval about $0 \in \mathbb{R}$, such that*

(i) $(d/ds)\Gamma(x, s) = V(\Gamma(x, s))$,

(ii) $\Gamma(x, s)$ *is once more continuously differentiable with respect to x than the components of V,*

(iii) $\Gamma(x, 0) = x$ *for all x.*

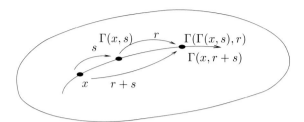

Figure 2.9 The 'flow equation', $\Gamma(\Gamma(x, s), r) = \Gamma(x, r + s)$.

Remark 2.3.9 Theorem 2.3.8 is the Existence and Uniqueness of Solutions of (first order) Ordinary Differential Equations (EUSODE) with a given initial value. The best proofs require only continuity of the components of V and use the Arzela–Ascoli Theorem, see for example Brown (1993), Theorem 1.1. The significance of this result, its utility and the depth of its proof, cannot be overestimated.

Theorem 2.3.10 *Integral curves for a vector field V satisfy the* flow equation,

$$\Gamma(\Gamma(x, s), r) = \Gamma(x, r + s),$$

that is, if you flow for time s, and then flow for time r, the result is the same as flowing for time $r + s$ (Figure 3.9). If the flows are only partial, we need to assume that $\Gamma(x, r + s)$ is defined.

Proof The flow equation follows from the fact that the two curves

$$\gamma_1 : r \mapsto \Gamma(\Gamma(x, s), r), \qquad \gamma_2 : r \mapsto \Gamma(x, s + r)$$

both satisfy the same differential equation and the same initial condition, and hence are the same curve:

$$\gamma_1(0) = \Gamma(\Gamma(x, s), 0) = \Gamma(x, s) = \gamma_2(0),$$

$$\frac{d}{dr}\gamma_1(r) = \frac{d}{dr}\Gamma(\Gamma(x, s), r) = V(\Gamma(\Gamma(x, s), r)) = V(\gamma_1(r)),$$

$$\frac{d}{dr}\gamma_2(r) = \frac{d}{dr}\Gamma(x, s + r) = V(\gamma(x, s + r)) = V(\gamma_2(r)). \qquad \square$$

The flow equation means that integral curves of vector fields on M define actions of the Lie group $(\mathbb{R}, +)$ on M. Partial flows define only a local group action. This does not help to find the integral curves of course, but it does mean that a rich source of actions is available for finding counterexamples to conjectures concerning how orbits of actions are embedded in M.

2.3.1 Integral curves in terms of the exponential of a vector field

If the vector field is analytic, then its flow can be understood in terms of a series expansion as follows. Writing the vector field $\mathbf{v}$ in operator notation, so that for any analytic function f, the functions $\mathbf{v}^0(f) = f$, $\mathbf{v}(f)$, $\mathbf{v}^2(f) = \mathbf{v}(\mathbf{v}(f))$ and so on are defined, we further define $\exp(t\mathbf{v})$ for $t \in \mathbb{R}$ to be the operator given by

$$\exp t\mathbf{v}(f) = \sum_{n=0}^{\infty} \frac{1}{n!} t^n \mathbf{v}^n(f). \tag{2.4}$$

Note the value $x \in M$ at which these expressions are evaluated is implicit. Assuming the series converges and can be differentiated term by term, it is straightforward to show that

$$\frac{d}{dt} \exp t\mathbf{v}(f) = \mathbf{v}\left(\exp t\mathbf{v}(f)\right),$$

since $\mathbf{v}$ is linear. Since further $\exp(0)(f) = f$, we suspect that $\exp t\mathbf{v}(f)$ is the value of f on the integral curve of the vector field of $\mathbf{v}$ at time t, with initial value being the implicit x. The next theorem gives the explicit relationship between $\exp t\mathbf{v}$ and $\Gamma_t^{\mathbf{v}}$.

Theorem 2.3.11 *The series* $\exp t\mathbf{v}(f)$ *is the Taylor series of* $f \circ \Gamma_t^{\mathbf{v}}$, *since*

$$\left.\frac{d^n}{dt^n}\right|_{t=0} f \circ \Gamma_t^{\mathbf{v}} = \mathbf{v}^n(f) \tag{2.5}$$

for all $n > 0$ *and for* $n = 0$ *we have* $\mathbf{v}^0(f) = f = f \circ \Gamma_0^{\mathbf{v}}$.

Taking f to be the coordinate functions x_i yields the integral curve in coordinates. By and large, it is often far easier to solve the differential system for the flow given in Theorem 2.3.8 than it is to sum the series into closed form. But for analytic vector fields, useful information for small t can be obtained.

Exercise 2.3.12 Show the series (2.4) converges and can be differentiated term by term, provided both f and the components of $\mathbf{v}$ are analytic. Hint: Taylor series for analytic functions converge and can be differentiated term by term.

Exercise 2.3.13 The constant vector field $\mathbf{v} = \partial_x$ defined on the real line has for its flow, $\Gamma_t^{\mathbf{v}}(x) = x + t$. Show that $\exp(t\mathbf{v})(f)$ is the Taylor series of $f(x + t)$ based at x.

Exercise 2.3.14 The linear vector field $\mathbf{v} = ax\partial_x$ defined on the real line has for its flow, $\Gamma_t^{\mathbf{v}}(x) = \exp(at)x$. Verify equation (2.5) for $n = 1, 2$ and 3 for arbitrary analytic f. Taking $f(x) = x$ show that the series is indeed the Taylor series for $\exp(at)x$ based at x.

Exercise 2.3.15 Prove Theorem 2.3.11.

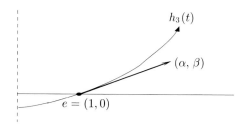

Figure 2.10 The curve $t \mapsto h_3(t)$ in G.

2.4 Tangent vectors at the identity versus one parameter subgroups

We can now state and prove the following major theorem.

Theorem 2.4.1 *Let G be a Lie group. There is a one-to-one correspondence between one parameter subgroups of G and tangent vectors at the identity $e \in G$.*

A one parameter subgroup $h(t) \subset G$ satisfies $h(0) = e$ and thus defines a path through e. Hence, these subgroups define tangent vectors in $T_e G$. It is the *converse* of the theorem that requires proof: given any $\mathbf{v} \in T_e G$, there is a unique one parameter subgroup $h_{\mathbf{v}}(t)$ whose tangent at e is $\mathbf{v}$. The map $\mathbf{v} \mapsto h_{\mathbf{v}}(t)$ is called the *exponential map* and is discussed in greater detail in the next section.

Example 2.4.2 Recall Exercise 1.5.4. For every tangent vector $\mathbf{v} = (\alpha, \beta)$ at the identity $e = (1, 0)$, there is a one parameter subgroup $h_{\mathbf{v}}(t)$ such that $h'_{\mathbf{v}}(0) = \mathbf{v}$. The path traced out by h_3 in G, and its corresponding tangent vector at $t = 0$ is depicted in Figure 2.10.

Note $(d/dt)|_{t=0} h_3(t) = (\alpha, \beta)$ is valid for all $\alpha \neq 0$, and $(d/dt)|_{t=0} h_1(t)) = (0, \beta)$. Thus, for *any* vector $\mathbf{v} \in \mathbb{R}^2$, we have a one parameter subgroup $h_{\mathbf{v}}(t)$ passing through e with tangent vector $\mathbf{v}$ at $t = 0$. The theorem proved in this section is that this subgroup is *unique*. Note: there are infinitely many paths through e with tangent vector (α, β). Only *one* of them is a one parameter subgroup.

Proof The proof of Theorem 2.4.1 hinges on the fact that left (or right) multiplication by h maps any vector $\mathbf{v} \in T_e G$ to a vector in $T_h G$. Indeed, given $[\gamma] \in T_e G$ then for any $h \in G$, we have $T L_h([\gamma]) = [h \cdot \gamma] \in T_h G$. Denote by

$$X_{[\gamma]} : G \longrightarrow TG, \qquad g \mapsto [g \cdot \gamma]$$

the vector field defined by left multiplication of the vector $[\gamma]$ (Figure 2.11).

Next, note the components of $X_{[\gamma]}(g)$ vary smoothly with g, i.e. are differentiable, because group multiplication is assumed to be smooth. Applying

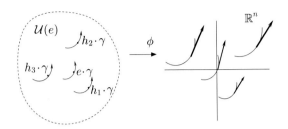

Figure 2.11 The vector field created by left multiplication of a vector at e.

Theorem 2.3.8, we obtain $\Gamma(e, s)$, the flow line of $X_{[\gamma]}$ satisfying $\Gamma(e, 0) = e$. Then we claim that

$$h(s) = \Gamma(e, s) \tag{2.6}$$

is a one parameter subgroup of G. Since $h(0) = e$ by construction, we need only to prove that

$$h(s) \cdot h(r) = h(s + r).$$

Let

$$\gamma(t) = \Gamma(e, t), \qquad |t| < \epsilon, \qquad \epsilon > 0$$

represent the tangent vector at e of the path $h(s)$. For any fixed s there are two paths at $h(s)$ which represent the vector $h'(s)$, namely $t \mapsto h(s) \cdot \gamma(t)$ and $t \mapsto \Gamma(\Gamma(e, t), s)$. Similarly for fixed r we have two paths representing the vector of the flowline at $h(r) \cdot h(s)$, namely $t \mapsto h(r) \cdot h(s) \gamma(t)$ and $t \mapsto \Gamma(\Gamma(\Gamma(e, t), s), r)$. But this last is equal to $t \mapsto \Gamma(\Gamma(e, t), s + r)$ which also represents the vector of the flowline at $h(r + s)$. Since vectors which are equal have the same basepoint, it must be that $h(r) \cdot h(s) = h(r + s)$.

Finally since a flowline through a point is *unique*, there is a *unique* one parameter subgroup for each $\mathbf{v} \in T_e G$. $\square$

2.5 The exponential map

Definition 2.5.1 The map $v \mapsto h_v$ which sends a vector in $T_e G$ to the one parameter subgroup h_v, constructed in the last section and which satisfies, in coordinates,

$$\frac{\mathrm{d}}{\mathrm{d}t}\bigg|_{t=0} h_v(t) = v$$

is called the *exponential* map.

If G is a matrix Lie group and $A \in T_e G$, then the vector field X_A constructed in the previous section is simply

$$X_A : G \to TG, \qquad X_A(g) = Ag,$$

since if $A = (\mathrm{d}/\mathrm{d}s)|_{s=0} \gamma(s)$, then $(\mathrm{d}/\mathrm{d}s)|_{s=0}(\gamma(s)g) = Ag$ as the product rule holds for matrix multiplication. The one parameter family associated to A, found in the previous section, solves the differential equation,

$$\frac{\mathrm{d}}{\mathrm{d}t} h(t) = A h(t), \qquad h(0) = I.$$

The solution is given formally as

$$h(t) = I + tA + \frac{1}{2} t^2 A^2 + \frac{1}{3!} t^3 A^3 + \cdots + \frac{1}{n!} t^n A^n + \cdots \tag{2.7}$$

and thus unsurprisingly, we write

$$h(t) \equiv \exp(tA).$$

Exercise 2.5.2 Show from the series expansion (2.7) that $\exp((r + s)A) = \exp(rA)\exp(sA)$. Hint: calculate

$$\left. \frac{\mathrm{d}}{\mathrm{d}t} \right|_{t=0} \exp((s - t)A) \exp(tA)$$

assuming term-by-term differentiation and the product rule hold.

Exercise 2.5.3 The matrices

$$f_1 = \begin{pmatrix} 0 & 1 \\ -1 & 0 \end{pmatrix}, \quad f_2 = \begin{pmatrix} i & 0 \\ 0 & -i \end{pmatrix}, \quad f_3 = \begin{pmatrix} 0 & i \\ i & 0 \end{pmatrix}$$

are all in $T_e SU(2)$. Find $\exp(t f_i)$, $i = 1, 2, 3$, and verify that each forms a one parameter subgroup of $SU(2)$.

The exponential map provides a map

$$\exp : T_e G \to G, \qquad \exp(v) = h_v(1),$$

also called, for better or worse, the exponential map. There exists $\delta > 0$ sufficiently small, such that $\exp$ is injective on the set of vectors with norm less than δ. This provides one means of obtaining coordinates for G centered at e.

2.6 Associated concepts for transformation groups

Given a transformation group $\mathcal{T}$, it is not always obvious which Lie group G lies behind it, in the sense that $\mathcal{T}$ is a presentation of G. If the transformations are only local, the problem of deciding, for example, what is the manifold of $\mathcal{T}$, is even worse. Fortunately, we can define concepts analogous to tangent vectors and the exponential map, staying within the conceptual framework of transformations on a space M.

Recall that a transformation group $\mathcal{T}$ is a group whose elements are invertible maps of a specified space M to itself. The action of $\mathcal{T}$ on M is evaluation; for $h \in \mathcal{T}$,

$$\mathcal{T} \times M \to M, \qquad (h, z) \mapsto h * z = h(z).$$

The group product is composition of mappings, and the identity element of $\mathcal{T}$ is the identity map on M, $\mathrm{id}|_M$. A path $h(t) \subset \mathcal{T}$ based at the identity, so that $h(0) = \mathrm{id}|_M$, yields for every $z \in M$ a path in M based at z, denoted variously as

$$h(t) * z = h(t)(z), \quad h_t(z) \quad \text{or} \quad h(t, z)$$

according to context to ease the exposition. Thus a path $h(t) \subset \mathcal{T}$ through the identity yields a vector field on M,

$$z \mapsto \left. \frac{\mathrm{d}}{\mathrm{d}t} \right|_{t=0} h(t)(z) \in T_z M.$$

If M has local coordinates $z = (z_1, \ldots, z_n)$, we have

$$\left. \frac{\mathrm{d}}{\mathrm{d}t} \right|_{t=0} h(t)(z_1, \ldots, z_n) = (\zeta_1(z), \ldots, \zeta_n(z))$$

for some functions ζ_i is the vector field in coordinates; these are the *infinitesimals* which are the components of the 'infinitesimal action' discussed in Section 1.6.

Definition 2.6.1 We say two paths in $\mathcal{T}$ based at the identity transformation are *equivalent* if the vector fields they generate on M are equal; in coordinates, if their infinitesimals are equal.

Conversely, given a vector field V on M, the flow map $z \mapsto \Gamma_t^V(z)$ yields a one parameter transformation group $h_V(t) = \Gamma_t^V$ on M. Indeed, applying Theorem 2.3.8, we have

$$h_V(s) \circ h_V(t) = \Gamma_t^V(\Gamma_s^V) = \Gamma_{t+s}^V = h_V(s + t)$$

by the flow condition. The flow may generate only a local one parameter transformation group, since $h_V(t)$ may not be defined for all t, only for t sufficiently close to 0. By the uniqueness of the flow maps given by Theorem 2.3.8, we have proved the following theorem, analogous to Theorem 2.4.1.

Theorem 2.6.2 *Equivalence classes of paths in the (local) Lie transformation group T based at the identity transformation are in one-to-one correspondence with (local) one parameter subgroups of T.*

An example of a one parameter flow obtained by integrating the infinitesimal vector field is given in Exercise 1.6.27.

Definition 2.6.3 The space of vector fields on M is denoted $\mathcal{X}(M)$. The subset of vector fields generated by the transformation group T is denoted $\mathcal{X}_T(M)$.

Theorem 2.6.4 *If M is a manifold and T a transformation group on M, then both $\mathcal{X}(M)$ and $\mathcal{X}_T(M)$ are vector spaces.*

Proof Vector fields are maps $\zeta : M \to T(M)$, and each $T_x M$ is a vector space. Thus the set of vector fields $\mathcal{X}(M)$ with addition and scalar multiplication defined point wise is also a vector space.

To show $\mathcal{X}_T(M)$ is a vector space, let $\zeta_1, \zeta_2 \in \mathcal{X}_T(M)$ be the infinitesimals for the paths $h_1(t), h_2(t) \subset T$ based at the identity map. Then $\zeta_1 + \zeta_2$ is the infinitesimal for the path $h_1(t) \circ h_2(t) \subset T$, while if $k \in \mathbb{R}$, then $k\zeta$ is the infinitesimal for the path $h(kt) \subset T$. $\qquad\qquad\square$

It is now hopefully clear that:

the object analogous to $T_e G$ for a transformation group T is $\mathcal{X}_T(M)$, since it is a vector space whose elements generate the one parameter subgroups of T.

We can now write down a rigorous definition of the infinitesimal action, given for one parameter Lie groups in Definition 1.6.1.

Definition 2.6.5 If the element $v \in T_e G$ is represented by a path $\gamma(t)$ in G based at the identity e, that is, $\gamma(0) = e$ and $\gamma'(0) = v$, then the group action $G \times M \to M$ defines a map

$$T_e G \to \mathcal{X}(M), \qquad v \mapsto X_v$$

where

$$X_v(x) = \frac{\mathrm{d}}{\mathrm{d}t}\Big|_{t=0} \gamma(t) \cdot x \in T_x M$$

and which we will call the *infinitesimal action* of G on M. The vector field X_v is called the *infinitesimal vector field* corresponding to the path $\gamma(t) \subset G$.

A similar definition holds for an element $v \in \mathcal{X}_T(M)$.

Note the infinitesimal action is not actually an action of either G or $T_e G$ on M. However, the flow of the so-called 'infinitesimal vector fields' gives the action.

Exercise 2.6.6 Show that X_v is well defined, that is, does not depend on which representative path γ for v is used.

If we apply the Theorem 2.6.2 to the action of a Lie group on itself, given by left multiplication, then we obtain Theorem 2.4.1. Using right multiplication leads to the same theorem, although the vector fields to integrate are different. More interesting is applying the above to the conjugation (or adjoint) action (see Exercise 1.2.10),

$$G \times G \to G, \qquad (g, h) \mapsto g^{-1} h g,$$

a typical example is considered in the next exercise.

Exercise 2.6.7 Suppose that G is a matrix group with identity I, so that the product rule for differentiation holds. Let $A(t)$ be a path in G based at I, and consider

$$B(t) = A(t)^{-1} B_0 A(t), \qquad B(0) = B_0.$$

Show

•
$$\frac{\mathrm{d}}{\mathrm{d}t} B(t) = [B, A(t)^{-1} A'(t)] \tag{2.8}$$

where $[P, Q] = PQ - QP$ is the usual matrix bracket, see also Example 1.6.4, and

• $A(t)^{-1} A'(t) \in T_I G$ for all t.

Bracket equations like equation (2.8) arise for example in mechanics, the most famous example being the spinning top. Thus, solving the system $B'(t) = [C(t), B(t)]$ where $B(t) \in G$ and $C(t) \in T_I G$, involves solving the simpler system $A'(t) = A(t)C(t)$ for $A(t)$; one then has $B(t) = A(t)^{-1} B_0 A(t)$.

For further reading on Lie groups, see Gilmore (1974) and Tapp (2005).

3

From Lie group to Lie algebra

In the previous chapter, we discussed the tangent structure on a Lie group, and the relationship between the set of one parameter subgroups and the tangent space T_eG at the identity element. The most striking feature of the tangent space at the identity of a Lie group is the existence of a natural product, called a *Lie bracket*, so that T_eG is an algebra; the Lie algebra of the Lie group.

Since Lie groups arise in different formulations, so does the appearance of the bracket in the Lie algebra. However, they all follow from the one formula for the Lie bracket of two vector fields on $\mathbb{R}^n$ which we consider in the first section. The geometric formulation looks unusable in practice, so we 'deconstruct' it to make it easily computable, prove some of its properties and discuss the all important Frobenius Theorem. We then derive the Lie algebra bracket for a general Lie group in Section 3.2, giving details in the two main cases of interest, matrix groups in Section 3.2.1 and transformation groups in Section 3.2.2. Although many authors simply give the formulae for the Lie bracket in these two cases as the definition of the Lie bracket, and readers only needing to compute can skip straight to these formulae, it is both interesting and helpful to know that in fact they are both instances of the same geometric construction. Since so many of the proofs are straightforward if one knows about the underlying construction and almost impossible if one does not, leaving out the basic underlying structure of the Lie bracket would have been counterproductive.

In Section 3.2.2 we discuss the so called Three Theorems of Lie, as originally formulated for transformation groups. Their historical significance is immense and they inform all and any understanding of a Lie group action. We then prove an important formula that will be central in proving the symbolic formulae for the differentiation of invariants in Chapter 4.

Finally in Section 3.3 we examine the Adjoint action of both matrix and transformation groups on their Lie algebras. A beautiful application of the latter will appear in Chapter 7 where we discuss variational methods and Noether's Theorem.

Notation[†]

A generic vector field on a manifold will be denoted X or Y.

An infinitesimal vector field will be denoted $\mathbf{v}$ or $\mathbf{w}$.

The infinitesimal vector field corresponding to the group parameter a will be denoted $\mathbf{v}_a$.

The vector field obtained by the action of a one parameter group h will be denoted $\mathbf{v}_h$.

An element of $T_e G$ will be denoted v or w.

A generic point in M or $\mathbb{R}^n$ will be denoted z. In many applications, z will denote the generic point $(x, u^\alpha, u_K^\alpha)$ in a jet bundle.

3.1 The Lie bracket of two vector fields on $\mathbb{R}^n$

Recall that if X is a vector field on $\mathbb{R}^n$, then the flowline or integral curve of X passing through z is denoted $t \mapsto \Gamma_t^X(z)$ and satisfies

$$\Gamma_0^X(z) = z, \qquad \frac{\mathrm{d}}{\mathrm{d}t}\Gamma_t^X(z) = X(\Gamma_t^X(z)),$$

and $\Gamma_t^X \Gamma_s^X = \Gamma_{t+s}^X$; see Theorem 2.3.10, and Figure 2.9.

Definition 3.1.1 If $\gamma(s)$ is a path in $\mathbb{R}^n$ based at z, so that

$$\gamma(0) = z, \qquad \gamma'(0) \in T_z\mathbb{R}^n,$$

then the *Jacobian* of the flow map Γ_t^X is

$$D\Gamma_t^X : T_z\mathbb{R}^n \to T_y\mathbb{R}^n, \qquad y = \Gamma_t^X(z)$$

and is defined by

$$D\Gamma_t^X(\gamma'(0)) = \frac{\mathrm{d}}{\mathrm{d}s}\bigg|_{s=0} \Gamma_t^X(\gamma(s)).$$

Thus, if a vector field $X = (X_1, X_2)$ on $\mathbb{R}^2$ has integral curves $t \mapsto \Gamma_t^X(z_1, z_2) = (F_1(z_1, z_2, t), F_2(z_1, z_2, t))$, then with vectors represented as

---

[†] It was not possible to please all my colleagues regarding notation for vectors and vector fields, since the notations in use are mutually exclusive. Some authors use X and Y for $g \cdot x$ and $g \cdot y$, others use X and Y for infinitesimals. Many use V for a vector field, while others reserve V to denote the scalar corresponding to potential energy, and instead use $\mathbf{f}$ for a field. The language used here is consistent with textbooks in Differential Topology and Olver (1993).

columns,

$$D\Gamma_t^X(\gamma'(0)) = \begin{pmatrix} \dfrac{\partial F_1}{\partial z_1} & \dfrac{\partial F_1}{\partial z_2} \\[2mm] \dfrac{\partial F_2}{\partial z_1} & \dfrac{\partial F_2}{\partial z_2} \end{pmatrix} \begin{pmatrix} \gamma_1'(0) \\ \gamma_2'(0) \end{pmatrix}.$$

Exercise 3.1.2 Using $\Gamma_{-t}^X \Gamma_t^X(z) = \Gamma_0^X(z) = z$, deduce using the chain rule

$$\left(D\Gamma_t^X\right)^{-1}\big|_z = D\Gamma_{-t}^X\big|_{\Gamma_t^X(z)}$$

We will also need the Jacobian of a vector field. In theory, this lies in $T(TM)$, the tangent bundle of the tangent bundle. In practice, we note that the derivative of a path of vectors in $\mathbb{R}^n$ can be identified with a vector in $\mathbb{R}^n$.

Definition 3.1.3 Given a vector field $X : M \to TM$ on M, and a path $s \mapsto \gamma(s)$ on M based at z, the *Jacobian* of the vector field X in the direction $\gamma'(0)$ is defined to be

$$DX(\gamma'(0)) = \frac{\mathrm{d}}{\mathrm{d}s}\bigg|_{s=0} X(\gamma(s)).$$

Example 3.1.4 Representing vectors as columns, the vector field $X = (X_1(z_1, z_2), X_2(z_1, z_2))^T$ on $\mathbb{R}^2$ has Jacobian in the direction $\gamma'(0) = (\gamma_1'(0), \gamma_2'(0))^T$ given by

$$DX(\gamma'(0)) = \frac{\mathrm{d}}{\mathrm{d}s}\bigg|_{s=0} \begin{pmatrix} X_1(\gamma(s)) \\ X_2(\gamma(s)) \end{pmatrix} = \begin{pmatrix} \dfrac{\partial X_1}{\partial z_1} & \dfrac{\partial X_1}{\partial z_2} \\[2mm] \dfrac{\partial X_2}{\partial z_1} & \dfrac{\partial X_2}{\partial z_2} \end{pmatrix} \begin{pmatrix} \gamma_1'(0) \\ \gamma_2'(0) \end{pmatrix}.$$

Exercise 3.1.5 Noting differentiation with respect to the flow parameter t and the coordinates z_i on M commute, and assuming X is a differentiable vector field, show

$$\frac{\mathrm{d}}{\mathrm{d}t}\bigg|_{t=0} D\Gamma_t^X(\gamma'(0)) = DX(\gamma'(0)).$$

Now let two vector fields X and Y be given, with integral curves $\Gamma_t^X(z)$ and $\Gamma_t^Y(z)$ respectively through a given point z. Consider Figure 3.1, which shows the integral curve $\Gamma_t^X(z)$ and the vectors $Y(z)$ and $Y(\Gamma_t^X(z))$. The map $(D\Gamma_t^X)^{-1}$ takes any vector at the point $\Gamma_t^X(z)$ to a vector at z. Define

$$V(t) = (D\Gamma_t^X)^{-1} Y(\Gamma_t^X(z)). \tag{3.1}$$

Then $V(t)$ is a path of vectors based at z, with $V(0) = Y(z)$. The Lie bracket of X and Y is, by definition, the derivative of $V(t)$ at $t = 0$.

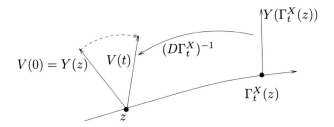

Figure 3.1 Construction of the Lie bracket of the vector fields X and Y.

Definition 3.1.6 (Lie bracket of vector fields) If X and Y are vector fields defined in a neighbourhood of z and X has an integral curve $t \mapsto \Gamma_t^X(z)$ through z, then the *Lie bracket* $[X, Y]$ at z is defined to be the vector

$$[X, Y](z) = \frac{\mathrm{d}}{\mathrm{d}t}\Big|_{t=0}(D\Gamma_t^X)^{-1}Y(\Gamma_t^X(z)). \tag{3.2}$$

The formula (3.2) requires serious 'deconstruction' before we can do anything with it. For example, it is not apparent that $[X, Y](z) = -[Y, X](z)$, although this turns out to be true. Also, it seems that we need to find the integral curves for X in order to calculate $[X, Y]$, which fortunately turns out to be false for differentiable vector fields.

The deconstruction process requires the vector fields to be at least once continuously differentiable, while the proofs of the properties of the bracket require higher orders of differentiability. To keep the exposition simple, we will assume that the vector fields we consider are smooth.

Theorem 3.1.7 *If X and Y are smooth vector fields, then*

$$[X, Y](z) = DY(X(z)) - DX(Y(z)). \tag{3.3}$$

Proof Define $V(t)$ as in equation (3.1). We seek $[X, Y] = V'(0)$. We have then

$$D\Gamma_t^X V(t) = Y(\Gamma_t^X(z)). \tag{3.4}$$

The left hand side is the product of a matrix $D\Gamma_t^X$ and a vector $V(t)$, while the right hand side is the composition of Y with Γ_t^X. Thus applying $\mathrm{d}/\mathrm{d}t$ to both sides, by the product rule on the left hand side and the chain rule on the right, we obtain

$$\left(\frac{\mathrm{d}}{\mathrm{d}t}D\Gamma_t^X\right)V(t) + D\Gamma_t^X V'(t) = DY(X(\Gamma_t^X(z))),$$

where we have used $(d/dt)\Gamma_t^X(z) = X(\Gamma_t^X(z))$. Since derivatives with respect to space and time coordinates commute, we have

$$\frac{d}{dt}D\Gamma_t^X = D\frac{d}{dt}\Gamma_t^X = D(X(\Gamma_t^X)) = (DX)(D\Gamma_t^X)$$

where the last equality follows from the chain rule. Thus,

$$D\Gamma_t^X V'(t) = DY(X(\Gamma_t^X(z))) - DX(Y(\Gamma_t^X(z))) \qquad (3.5)$$

where we have used equation (3.4) to simplify the final term. At $t = 0$, Γ_t^X is the identity map, and thus so is its Jacobian map, while by definition, $V'(0) = [X, Y](z)$. Hence, setting $t = 0$ in equation (3.5) yields the result. $\qquad\qquad\square$

First let us calculate equation (3.3) in coordinates, representing X and Y as column vectors. If

$$X(z) = \begin{pmatrix} f_1(z) \\ \vdots \\ f_n(z) \end{pmatrix}, \qquad Y(z) = \begin{pmatrix} g_1(z) \\ \vdots \\ g_n(z) \end{pmatrix} \qquad (3.6)$$

and if $\gamma(s)$ is a path based at z so that $\gamma'(0) = (z_1'(0), \ldots, z_n'(0))^T$ is a vector at z, then

$$DX(\gamma'(0)) = \frac{d}{ds}\Big|_{s=0} X(\gamma(s))$$

$$= \begin{pmatrix} \sum_j \dfrac{\partial f_1}{\partial z_j} z_j'(0) \\ \vdots \\ \sum_j \dfrac{\partial f_n}{\partial z_j} z_j'(0) \end{pmatrix}$$

$$= \begin{pmatrix} \dfrac{\partial f_1}{\partial z_1} & \cdots & \dfrac{\partial f_1}{\partial z_n} \\ \vdots & \ddots & \vdots \\ \dfrac{\partial f_n}{\partial z_1} & \cdots & \dfrac{\partial f_n}{\partial z_n} \end{pmatrix} \begin{pmatrix} z_1'(0) \\ \vdots \\ z_n'(0) \end{pmatrix}$$

and hence the Lie bracket

$$[X, Y](z) =$$

$$
\begin{pmatrix}
\dfrac{\partial g_1}{\partial z_1} & \cdots & \dfrac{\partial g_1}{\partial z_n} \\
\vdots & \ddots & \vdots \\
\dfrac{\partial g_n}{\partial z_1} & \cdots & \dfrac{\partial g_n}{\partial z_n}
\end{pmatrix}
\begin{pmatrix} f_1 \\ \vdots \\ f_n \end{pmatrix}
-
\begin{pmatrix}
\dfrac{\partial f_1}{\partial z_1} & \cdots & \dfrac{\partial f_1}{\partial z_n} \\
\vdots & \ddots & \vdots \\
\dfrac{\partial f_n}{\partial z_1} & \cdots & \dfrac{\partial f_n}{\partial z_n}
\end{pmatrix}
\begin{pmatrix} g_1 \\ \vdots \\ g_n \end{pmatrix}. \tag{3.7}
$$

Exercise 3.1.8 Given two vector fields on $\mathbb{R}^3$ with coordinates (x, y, z) expressed as column vectors, $X = (y^2 + x, z, x^3)^T$ and $Y = (9z^2, 2x + 1, x^2 + y^2)^T$, show that

$$
DX = \begin{pmatrix} 1 & 2y & 0 \\ 0 & 0 & 1 \\ 3x^2 & 0 & 0 \end{pmatrix}, \qquad
DY = \begin{pmatrix} 0 & 0 & 18z \\ 2 & 0 & 0 \\ 2x & 2y & 0 \end{pmatrix}
$$

and hence that $[X, Y]$ equals

$$
(18x^3z - 9z^2 - 2y(2x + 1), 2(x + y^2) - x^2 - y^2,
$$
$$
2x(x + y^2) + 2yz - 27x^2z^2)^T.
$$

Definition 3.1.9 Representing the vector fields X and Y in (3.6) as *operators*,

$$
X = \sum f_i \frac{\partial}{\partial z_i}, \qquad Y = \sum g_i \frac{\partial}{\partial z_i},
$$

define the product XY to be the operator $(XY)(\phi) = X(Y(\phi))$ for any sufficiently smooth function ϕ (we assume the domains of ϕ, X and Y are such that $X(Y(\phi))$ makes sense).

It is a straightforward calculation to check that the right hand side of equation (3.7), in operator form, is given by

$$[X, Y] = XY - YX \tag{3.8}$$

since the second order derivative terms always cancel.

Exercise 3.1.10 Expressing the vector fields in Exercise 3.1.8 as operators, so that $X = (y^2 + x)\partial_x + z\partial_y + x^3\partial_z$ and $Y = 9z^2\partial_x + (2x + 1)\partial_y + (x^2 + y^2)\partial_z$, verify

$$
XY - YX = (18x^3z - 9z^2 - 2y(2x + 1))\partial_x
$$
$$
+ (2(x + y^2) - x^2 - y^2)\partial_y
$$
$$
+ (2x(x + y^2) + 2yz - 27x^2z^2)\partial_z
$$

which is $[X, Y]$ in operator form.

The formula (3.8) allows easy proofs of the following properties of the Lie bracket of vector fields.

Theorem 3.1.11 *The Lie bracket of smooth vector fields*

(i) *is skew symmetric,* $[X, Y] = -[Y, X]$,
(ii) *is bilinear,* $[aX_1 + bX_2, Y] = a[X_1, Y] + b[X_2, Y]$ *where* $a, b \in \mathbb{R}$,
(iii) *satisfies the Jacobi identity,*

$$[X, [Y, Z]] + [Y, [Z, X]] + [Z, [X, Y]] = 0. \qquad (3.9)$$

Moreover,

(iv) *if f and g are differentiable functions from the domain of the vector fields X and Y, to $\mathbb{R}$, then*

$$[fX, gY] = fg[X, Y] + fX(g)Y - gY(f)X, \qquad (3.10)$$

where $(fX)(z) = f(z)X(z)$.

Exercise 3.1.12 Prove Theorem 3.1.11.

Definition 3.1.13 Suppose there exists a differentiable function f from the domain of the vector field X to the domain of vector field X', such that

$$X'(f(z)) = Df(X(z)).$$

Then we say that X and X' are f-related.

We showed in Section 2.2.2 how to calculate $Df(X(z))$ in coordinates. It is not the case that the image of a vector field under a Jacobian map is always a vector field; if $f(z^1) = f(z^2)$ then it is necessary that $Df(X(z^1)) = Df(X(z^2))$, otherwise $Df(X)$ is not well defined (is multi-valued).

Theorem 3.1.14 *Suppose f is a differentiable function from the domain of the vector fields X and Y to the domain of vector fields X' and Y' such that both*

$$X'(f(z)) = Df(X(z)), \qquad Y'(f(z)) = Df(Y(z)).$$

Then

$$[X', Y'](f(z)) = Df([X, Y])(z). \qquad (3.11)$$

In words, if X and X' are f-related, and Y and Y' are f-related, then $[X, Y]$ and $[X', Y']$ are f-related.

Proof To differentiate $X'(f(z)) = Df(X(z))$ in the direction of Y, we use the chain rule on the left and the product rule on the right, as Df is a matrix. We

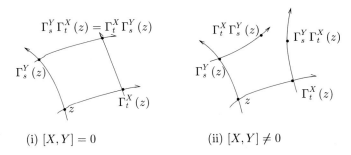

(i) $[X, Y] = 0$ (ii) $[X, Y] \neq 0$

Figure 3.2 Flows commute if and only if their Lie brackets are zero, Theorem 3.1.16.

then obtain

$$DX'(Df(Y)) = D^2 f(X, Y) + Df(DX(Y))) \qquad (3.12)$$

where $D^2 f(X, Y)$ in coordinates is a vector whose ith component is

$$Y^T \mathrm{Hessian}(f^i)X = \sum_{k,\ell} \frac{\partial^2 f^i}{\partial z^k \partial z^\ell} X^k Y^\ell$$

where $f^i(z)$ is the ith component of $f(z)$. It can be seen that $D^2 f(X, Y) = D^2 f(Y, X)$ as the partial derivatives commute. Similarly,

$$DY'(Df(X)) = D^2 f(X, Y) + Df(DY(X))). \qquad (3.13)$$

Subtracting equation (3.13) from (3.12) yields

$$DX'(Y') - DY'(X') = Df(DX(Y) - DY(X)),$$

the desired result.   □

Example 3.1.15 Let $\mathbb{R}$ have the coordinate (x), and set $X = \partial_x$ and $Y = x \partial_x$ to be two vector fields on $\mathbb{R}$. Let $f : \mathbb{R} \to \mathbb{R}$ be the exponential map, $f(x) = y = \exp(x)$. Then $[X, Y] = X$, and using Exercise 2.2.17, $X'(y) = y\partial_y$, $Y'(y) = y \log(y)\partial_y$ and

$$[X', Y'](y) = X'(y) = Df(X(x)) = Df([X, Y](x))$$

as required.

Theorem 3.1.16 *Given vector fields X and Y, then $[X, Y] = 0$ if and only if $\Gamma_t^X \Gamma_s^Y = \Gamma_s^Y \Gamma_t^X$. In words, the bracket of X and Y is identically zero if and only if the flows they define commute; see Figure 3.2.*

Proof First assume $[X, Y] \equiv 0$. Then $\left(D\Gamma_t^X\right)^{-1}\left(Y(\Gamma_t^X(z))\right)$ is a constant, which can be obtained by its value at $t = 0$. Now $\Gamma_0^X(z) = z$ for all z, so

$D\Gamma_0^X$ is the identity map, and thus

$$Y\big(\Gamma_t^X(z)\big) = D\Gamma_t^X\big(Y(z)\big)$$

for all z and t. Now consider the map,

$$G : (s, t) \mapsto \Gamma_s^X\big(\Gamma_t^Y(z)\big). \qquad (3.14)$$

Then

$$\begin{aligned}
\frac{\partial G}{\partial t}\bigg|_{t=t_0} &= D\Gamma_s^X Y\big(\Gamma_{t_0}^Y(z)\big) \\
&= Y\big(\Gamma_s^X \Gamma_{t_0}^Y(z)\big) \\
&= Y(G(s, t_0)).
\end{aligned}$$

So for fixed s, $G(s, t)$ satisfies the definition of the integral curves for Y, namely,

$$\frac{\mathrm{d}}{\mathrm{d}t} G(s, t) = Y\big(G(s, t)\big)$$

with initial condition

$$G(s, 0) = \Gamma_s^X(z).$$

Thus,

$$G(s, t) = \Gamma_t^Y\big(\Gamma_s^X(z)\big). \qquad (3.15)$$

Comparing (3.14) and (3.15) we have proved the flows commute.

Conversely, assume $\Gamma_t^Y \Gamma_s^X = \Gamma_s^X \Gamma_t^Y$. Differentiating both sides with respect to s at $s = 0$ yields $Y(\Gamma_t^X(z)) = D\Gamma_t^X(Y(z))$ so that

$$\big(D\Gamma_t^X\big)^{-1} Y\big(\Gamma_t^X(z)\big) = Y(z).$$

The right hand side is independent of t, so that differentiating again with respect to t yields $[X, Y] \equiv 0$. $\qquad\square$

We will use the result in the following exercise in Section 4.6.

Exercise 3.1.17 Show that $[X, Y] = -Y$ if and only if

$$\Gamma_s^X \Gamma_{\exp(-s)t}^Y = \Gamma_t^Y \Gamma_s^X.$$

Hint: adapt the proof of Theorem 3.1.16. Generalise the result to the case $[X, Y] = \lambda Y$.

We can extend Theorem 3.1.16 to sets of pairwise commuting vector fields, which we assume to be linearly independent. Thus, suppose $X_1, X_2, \dots, X_k$

are pointwise linearly independent vector fields defined near z with

$$[X_i, X_j] = 0, \qquad i, j = 1, \ldots, k.$$

Define

$$\phi : (t_1, \ldots, t_k) \mapsto \Gamma^{X_1}_{t_1} \cdots \Gamma^{X_k}_{t_k}(z),$$

with $\phi(0) = z$. Then, by the commutativity of the flows, we have

$$\frac{\partial \phi}{\partial t_j} = X_j(\phi(t)).$$

Hence ϕ represents an invertible transformation from a neighbourhood of $0 \in \mathbb{R}^k$ to a surface in $\mathbb{R}^n$ containing z. On this surface, we can use $(t_1, \ldots, t_k)$ as coordinates, and in these coordinates,

$$X_i \equiv \frac{\partial}{\partial t_i}.$$

The surface is called an *integral element* of the vector fields X_i.

It is not necessary for vector fields to pairwise commute for an integral element to exist. The necessary and sufficient condition is given in the Frobenius Theorem which we discuss next.

3.1.1 Frobenius' Theorem

This theorem is a key ingredient in understanding how the orbits of a group action foliate the space on which the group acts, so we investigate it in detail. We assume linearly independent, and in particular non-zero, vector fields, but not just for simplicity, it is the case we need.

Definition 3.1.18 Suppose $X_1, \ldots, X_k$ are linearly independent vector fields on $U \subset \mathbb{R}^n$ such that for each i, j there exist functions $c^\ell_{ij} : U \to \mathbb{R}$ so that

$$[X_i, X_j](z) = \sum_{\ell=1}^k c^\ell_{ij}(z) X_\ell(z). \tag{3.16}$$

Then we say the vector fields $X_1, \ldots, X_k$ are in *involution*.

Pairwise commuting vector fields are in involution, as are their linear combinations with variable coefficients, by Theorem 3.1.11 (iv).

Exercise 3.1.19 In $\mathbb{R}^3$ with coordinates (x, y, z), show that $X_1 = y\partial_x - x\partial_y$, $X_2 = z\partial_x - x\partial_y$ are *not* in involution.

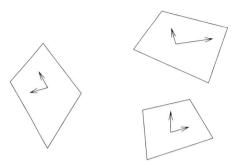

Figure 3.3 Three elements of a 2-plane field in $\mathbb{R}^3$.

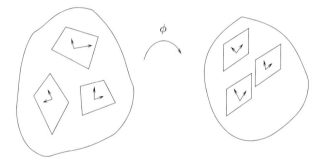

Figure 3.4 The image of a k-plane field under a foliation chart map ϕ.

Definition 3.1.20 Let $X_1, \ldots, X_k$ be linearly independent vector fields defined on $U \subset \mathbb{R}^n$. Let $P(z) \subset T_z M$ be the k dimensional plane spanned by the X_i at z. Then the set

$$P = \{P(z) \,|\, z \in U\} \subset T_U M$$

is called the k-*plane field* generated by the X_i, illustrated in Figure 3.3. If the X_i are in involution, we say P is *involutive*, and by virtue of Frobenius' Theorem below, we also say P is *integrable*.

Frobenius' Theorem provides a 'trivialisation' of an involutive k-plane field in the form of a foliation chart map. The foliation map sends the k-plane field P to the trivial k-plane field tangent to the parallel planes, $\mathbb{R}^k \times (c_{m+1}, \ldots, c_n)$ where the c_i are constants, see Figure 3.4. In other words, there is a change of coordinates such that in the new coordinates one can take for a basis of the plane field the vectors $\partial_i = \partial/\partial t_i$, for $i = 1, \ldots, k$.

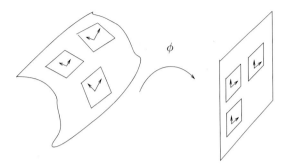

Figure 3.5 The preimage of a coordinate k-plane under the foliation map ϕ is a surface to which the k-plane field is tangent.

Definition 3.1.21 The coordinates $(t_1, t_2, \ldots, t_k)$ such that a basis for a given involutive k-plane field is $\partial_i = \partial/\partial t_i$, for $i = 1, \ldots, k$, are called *canonical coordinates* for that plane field.

Remark 3.1.22 Classical methods to solve invariant ordinary differential equations often relied on finding these coordinates. The use of moving frames makes this difficult step unnecessary. Canonical coordinates are important from the theoretical point of view. They give a clear picture of how group orbits foliate a space, locally, near a point where the action is non-singular.

The preimage under ϕ, of the parallel planes giving the k-plane field in the canonical coordinates, are the *integral elements* of the vector fields X_i, that is, are surfaces to which the X_i are tangent, see Figure 3.5. These surfaces can then be seen to foliate the space in the same way that parallel planes foliate $\mathbb{R}^n$.

Exercise 3.1.23 Consider $\mathbb{R}^3 \setminus \{0\}$, with coordinates (x, y, z). Let X and Y be two linearly independent vector fields that are orthogonal to the radial vector field $\mathbf{r} = x\partial_x + y\partial_y + z\partial_z$. Show that X, Y are in involution. What are the integral surfaces? Hint: show $DX(Y) - DY(X)$ is also orthogonal to $\mathbf{r}$.

Theorem 3.1.24 (Frobenius' Theorem) *Suppose that the vector fields $X_1, \ldots, X_k$ are in involution in U. Let P be the k-plane field defined by the X_i. Then for each $z \in U$ there exists a neighbourhood $U(z) \subset U$ about z, and a diffeomorphism $\phi : U(z) \to \mathbb{R}^n$ such that $\{P(y) \mid y \in U(z)\}$, the k-plane field generated by the X_i, is given by*

$$P(y) = (T\phi)^{-1}\big|_{\phi(y)}(\mathbb{R}^k \times \{0\}) \qquad (3.17)$$

for all $y \in U(z)$. The map ϕ is called a foliation chart *for P.*

Flowlines for X

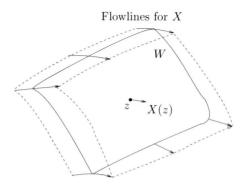

Figure 3.6 Diagram for claim 2, proof of Frobenius' Theorem.

Proof Suppose such a diffeomorphism exists. Define

$$Y_i = (T\phi)^{-1}\big|_{\phi(y)}\frac{\partial}{\partial z_i}, \qquad i = 1, \ldots, k$$

where $(z_1, z_2, \ldots, z_n)$ are the coordinates on $\mathbb{R}^n$. Then by Theorem 3.1.14, we have $[Y_i, Y_j] = 0$. Since the X_i are linear combinations of the Y_i, the involutivity condition follows from Theorem 3.1.11 (iv).

The converse is proved by induction on k. If $k = 0$, there is nothing to prove. If $k = 1$, then the result follows from the existence and uniqueness of integral curves of vector fields, Theorem 2.3.8, see Exercise 3.1.25.

Let Γ_t be the flow map for X_1.

Claim 1 The flow Γ_t determined by X_1 preserves the set of k-planes $\{P(z) \mid z \in U\}$,

$$T\Gamma_t P(z) = P(\Gamma_t(z)) \tag{3.18}$$

where $T\Gamma_t$ is the tangent map to Γ_t (recall in coordinates $T\Gamma_t$ is $D\Gamma_t$, the Jacobian of Γ_t), see Exercise 3.1.26.

Claim 2 Let W be an $n - 1$ dimensional surface through z such that $\Gamma : (-\epsilon, \epsilon) \times W \to U(z)$, $(s, y) \mapsto \Gamma_s(y)$, is a diffeomorphism onto a neighbourhood $U(z)$ of z, see Figure 3.6. Let $\bar{P}(y) = P(y) \cap T_y W$. Then any set of vector fields spanning the set of planes $\{\bar{P}(y)\}$ satisfies the involutivity condition, equation (3.16), see Exercise 3.1.27.

Claim 3 The foliation chart ψ for $\{\bar{P}(y)\}$ on W, which exists by the inductive step, can be extended to a foliation chart on $U(z) \approx (-\epsilon, \epsilon) \times W$. Indeed, set $\phi(\Gamma_t y) = (t, \psi(y))$, see Exercise 3.1.28. $\square$

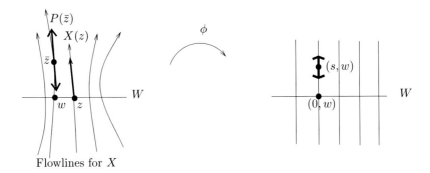

Figure 3.7 Diagram for Exercise 3.1.25. The 1-plane field is the set of tangent spaces to the flowlines.

Exercise 3.1.25 Prove the existence of a foliation chart in the case of a single non-zero vector field X. Hint: let W be a plane through z such that $X(z) \notin T_z W$. Then any point $\bar{z}$ in a neighbourhood of z is of the form $x = \Gamma_s(w)$ for some $w \in W$ and some $s \in \mathbb{R}$. Consider the map $\phi(\Gamma_s^X(w)) = (s, w)$ taking a neighbourhood $U(z)$ onto $(-\epsilon, \epsilon) \times W$ for some $\epsilon > 0$, see Figure 3.7. We have $P(\bar{z}) = \mathbb{R} X(\bar{z})$ and $T_\phi(P(\bar{z})) = \mathbb{R} \times \{0\} \subset T_{(s,w)}((-\epsilon, \epsilon) \times W)$. The fact that ϕ is a foliation chart, that is, satisfies equation (3.17), follows from

$$\mathbb{R} \cdot \frac{d}{ds}\phi(\Gamma_s^X(w)) = \mathbb{R} \cdot T\phi(X(\Gamma_s^X(w))) = T\phi(P(\bar{z})). \tag{3.19}$$

Exercise 3.1.26 Let $X_1, \ldots, X_n$ satisfy the involutivity condition, equation (3.16), let $P(z)$ be the k-plane spanned by the X_i at z, and let Γ_t be the flow map of X_1. Let

$$u_i(t) = D\Gamma_t(X_i(\Gamma_t(z))).$$

Then $u_i(0) = X_i(z)$. Use Theorem 3.11 to show that

$$\frac{d}{dt}\Big|_{t=s} u_i = D\Gamma_s([X_1, X_i](\Gamma_s(x))).$$

Hence show that u_i satisfies a first order linear system of differential equations, whose solution exists by the Fundamental Existence and Uniqueness Theorem for ordinary differential equations. Show that $u_i(t) \in P(\Gamma_t(z))$ follows from the uniqueness of the solution and the initial conditions, and hence prove Claim 1.

Exercise 3.1.27 Prove Claim 2. Hint 1: it suffices to find one set of vector fields spanning the set $\{\bar{P}(y) \mid y \in U(z)\}$ satisfying the involutivity condition. Hint 2: set $X_i'(y) = X_i - \alpha_i(y)X_1$ for some function $\alpha_i : U(z) \to \mathbb{R}$.

Exercise 3.1.28 Prove Claim 3. Hint: adapt the proof from Exercise 3.1.25, but using equation (3.18) proved in Claim 1 in (3.19) in place of the one parameter flow condition.

3.2 The Lie algebra bracket on T_eG

In the previous section, we defined the Lie bracket of vector fields on $\mathbb{R}^n$. Here we consider the Lie bracket of vector fields on a manifold, specifically, our Lie Group G, which is used to define a Lie bracket on T_eG. We then consider the details for matrix Lie groups and transformation groups.

Exercise 3.2.1 Let X and Y be vector fields defined on a manifold M. In each coordinate patch, $[X, Y]$ can be defined using the method of the previous section. Show that $[X, Y]$ is well defined as a vector field on the whole of M. Hint: Use Theorem 3.1.14 with $f = \phi_i \circ \phi_j^{-1}$.

Definition 3.2.2 Given $v, w \in T_eG$, the *Lie bracket* $[v, w] \in T_eG$ is obtained as follows.

Step 1 Extend v and w to vector fields $\widehat{v}, \widehat{w}$ on all of G by left multiplication: if $v = (\mathrm{d}/\mathrm{d}_s)|_{s=0}\gamma(s) \in T_eG$ then

$$\widehat{v}(g) = \frac{\mathrm{d}}{\mathrm{d}s}\bigg|_{s=0} g \cdot \gamma(s) \in T_gG,$$

and similarly for $\widehat{w}$.
Step 2 Calculate $[\widehat{v}, \widehat{w}]$.
Step 3 Define

$$[v, w] = [\widehat{v}, \widehat{w}](e) \in T_eG, \tag{3.20}$$

that is, the Lie bracket of the vector fields $\widehat{v}, \widehat{w}$ evaluated at the identity element.

Theorem 3.2.3 *The Lie bracket on T_eG is skew symmetric, bilinear, and satisfies the Jacobi identity.*

Proof The first two properties follow trivially from that of the Lie bracket of vector fields. In order to prove the Jacobi identity, we note that by construction, the vector field $\widehat{v}$ constructed in Step 1 above is L_g-related to itself for all $g \in G$, where $L_g : G \to G$ is left multiplication by g. By Theorem 3.1.14, we have

$$\widehat{[v, w]} = [\widehat{v}, \widehat{w}]. \tag{3.21}$$

Since for v, w, $r \in T_eG$ we have $[\hat{v}, [\widehat{w, r}]] + [\hat{w}, [\widehat{r, v}]] + [\hat{r}, [\widehat{v, w}]] = 0$, applying (3.21) yields $[\hat{v}, \widehat{[w, r]}] + [\hat{w}, \widehat{[r, v]}] + [\hat{r}, \widehat{[v, w]}] = 0$. Evaluating this last at the identity element yields the result.   □

Remark 3.2.4 Note that the Lie algebra bracket is *not* associative. The Jacobi identity replaces associativity in any calculation.

Definition 3.2.5 The *Lie algebra* $\mathfrak{g}$ associated to a Lie group G is defined to be T_eG, the vector space of tangent vectors based at the identity element of G, together with the bracket $\mathfrak{g} \times \mathfrak{g} \to \mathfrak{g}$, $(v, w) \mapsto [v, w]$ defined in (3.20).

The next definition collects standard names for some classical Lie algebras.

Definition 3.2.6 The Lie algebra associated with $GL(n)$ is denoted $\mathfrak{gl}(n)$, the Lie algebra associated with $SL(n)$ is denoted $\mathfrak{sl}(n)$, the Lie algebra associated with $SO(n)$ is denoted $\mathfrak{so}(n)$, and so forth. We denote by $\mathfrak{gl}(V)$ the set of all linear maps from the vector space V to itself.

In the next section, we will prove that $\mathfrak{gl}(n, \mathbb{F}) = M_n(\mathbb{F})$, the set of all linear maps from $\mathbb{F}^n$ to itself.

More generally, an algebra L whose product is bilinear, skew symmetric and satisfies the Jacobi identity, is called a Lie algebra, in which case the product is called a bracket, even when there is no apparent Lie group G for which $L = T_eG$.

We give now some important definitions of the *Ad*, *ad* actions, and also for Lie algebra homomorphisms and representations. These will be illustrated in the next two subsections, where we discuss the details of the Lie bracket for matrix Lie algebras and the concepts analogous to Lie algebra and Lie bracket for transformation groups.

There is an important action of G on $\mathfrak{g}$ induced by the conjugation action, $G \times G \to G$, $(g, h) \mapsto g^{-1}hg$.

Definition 3.2.7 Let $\gamma(t)$ be a path based at e so that $[\gamma] \in T_eG = \mathfrak{g}$. Then define the *Adjoint* action Ad of G on $\mathfrak{g}$ by

$$Ad : G \times \mathfrak{g} \to \mathfrak{g}, \qquad [\gamma] \mapsto [g^{-1}\gamma g]. \tag{3.22}$$

Remark 3.2.8 The conjugation action is also called the adjoint action of G on itself.

Further details of the *Ad* action for the specific case of matrix Lie algebras is given in equation (3.26), and for transformation groups in Definition 3.3.1.

Exercise 3.2.9

(i) Show Ad is well defined, that is, if in coordinates $\gamma_1'(0) = \gamma_2'(0)$ then

$$\frac{d}{dt}\bigg|_{t=0} g^{-1}\gamma_1(t)g = \frac{d}{dt}\bigg|_{t=0} g^{-1}\gamma_2(t)g.$$

(ii) Show Ad is a right action.

(iii) Show $Ad_g : \mathfrak{g} \to \mathfrak{g}$, given by $Ad_g(v) = Ad(g, v)$ is a linear map.

(iv) Show $g \cdot [x, y] = [g \cdot x, g \cdot y]$. Hint: use Theorem 3.1.14.

Definition 3.2.10 If $\mathfrak{g}$, $\mathfrak{h}$ are Lie algebras with Lie brackets $[\ ,\ ]_\mathfrak{g}$ and $[\ ,\ ]_\mathfrak{h}$ respectively, and $\phi : \mathfrak{g} \to \mathfrak{h}$ is a linear map, we say that ϕ is a *Lie algebra homomorphism* if for all $x, y \in \mathfrak{g}$,

$$\phi([x, y]_\mathfrak{g}) = [\phi(x), \phi(y)]_\mathfrak{h}.$$

If in addition $\mathfrak{h} = \mathfrak{gl}(V)$ where V is a vector space and $[A, B]_\mathfrak{h} = AB - BA$, the usual matrix Lie bracket, we say that ϕ is a *representation* of $\mathfrak{g}$. A Lie algebra homomorphism ϕ is said to be an *isomorphism* if ϕ is a bijection. A Lie algebra representation is said to be *faithful* if it is injective.

Since $\mathfrak{g}$ is a vector space, the set $\mathfrak{gl}(\mathfrak{g})$ of linear maps $\mathfrak{g} \to \mathfrak{g}$ is well defined. The Ad action induces a representation of $\mathfrak{g}$ in $\mathfrak{gl}(\mathfrak{g})$, also called the *adjoint representation*.

Definition 3.2.11 Let $v \in \mathfrak{g} = T_e G$ be represented as a path $\gamma(t)$ through $e \in G$, and let $w \in T_e G$. Then we define the adjoint representation of $\mathfrak{g}$ in $\mathfrak{gl}(\mathfrak{g})$ to be

$$ad_v : \mathfrak{g} \to \mathfrak{g}, \qquad ad_v(w) = \frac{d}{dt}\bigg|_{t=0} Ad_{\gamma(t)}(w).$$

Exercise 3.2.12 Show ad_v is well defined. Hint: apply standard arguments from multivariate calculus in coordinates.

Theorem 3.2.13 *For $v, w \in \mathfrak{g} = T_e G$,*

$$ad_v(w) = [v, w]. \qquad (3.23)$$

Proof Let $v, w \in \mathfrak{g} = T_e G$ be extended to vector fields $\widehat{v}, \widehat{w}$ on G as in the definition of the Lie bracket, so that $[v, w] = [\widehat{v}, \widehat{w}](e)$. Then representing v by $(d/dt)|_{t=0}\, \Gamma_t^{\widehat{v}}$, we have

$$ad_v(w) = \frac{d}{dt}\bigg|_{t=0} (D\Gamma_t^{\widehat{v}})^{-1}\widehat{w}(\Gamma_t^{\widehat{v}}(e)) = [v, w]. \qquad \square$$

Remark 3.2.14 For an arbitrary Lie algebra, not necessarily equal to T_eG for some Lie group G, equation (3.23) is taken as the definition of the adjoint action of the Lie algebra on itself.

Exercise 3.2.15 Prove that $ad : \mathfrak{g} \to \mathfrak{gl}(\mathfrak{g})$, $ad(x) \mapsto ad_x$, is a linear map, that is, $ad(x + y) = ad_{x+y} = ad_x + ad_y = ad(x) + ad(y)$, and $ad(\lambda x) = ad_{\lambda x} = \lambda ad_x = \lambda ad(x)$. Show further that ad is a representation of $\mathfrak{g}$ by showing that

$$ad([x, y]) = ad_{[x,y]} = ad_x ad_y - ad_y ad_x = ad(x) \circ ad(y) - ad(y) \circ ad(x).$$

3.2.1 The Lie algebra bracket for matrix Lie groups

Theorem 3.2.16 *If G is a matrix group and A, $B \in \mathfrak{g}$, then*

$$[A, B] = AB - BA,$$

the standard matrix bracket.

Proof For $A, B \in T_eG$, extending these to vector fields $\widehat{A}$, $\widehat{B}$ on all of G by left multiplication yields

$$\widehat{A}(g) = gA, \qquad \widehat{B}(g) = gB$$

where the products gA, gB are obtained by the standard multiplication of matrices. Recalling the discussion in Section 2.5 (where $\widehat{A}$ was denoted X_A, we have changed the notation to ease the exposition), we have

$$\Gamma_t^{\widehat{A}}(g) = g \exp(tA)$$

and thus

$$\widehat{B}(\Gamma_t^{\widehat{A}}(g)) = g \exp(tA)B.$$

Now

$$D\Gamma_t^{\widehat{A}}(\gamma'(0)) = \frac{d}{ds}\bigg|_{s=0} \Gamma_t^{\widehat{A}}(\gamma(s)) = \gamma'(0) \exp(tA)$$

and thus the inverse of $D\Gamma_t^{\widehat{A}}$ is right multiplication by $\exp(-tA)$. Hence

$$(D\Gamma_t^{\widehat{A}})^{-1}\widehat{B}(\Gamma_t^{\widehat{A}}(e))) = \exp(tA)B \exp(-tA).$$

Finally, we have

$$[A, B] = \frac{d}{dt}\bigg|_{t=0} \exp(tA)B \exp(-tA) = AB - BA$$

as required. $\qquad \square$

There is an enormous literature on matrix Lie algebras, solvable, nilpotent and semi-simple Lie algebras, the classification of semi-simple Lie algebras, their representations, root spaces, and so forth. The examples we describe here will be used in the later chapters, and give an idea of the kinds of calculations we will be needing.

Exercise 3.2.17 Consider the group $G = SL(2)$ of 2×2 matrices with determinant 1. Show that an arbitrary element in a neighbourhood of the identity is

$$g(a, b, c) = \begin{pmatrix} a & b \\ c & \dfrac{1 + bc}{a} \end{pmatrix}.$$

By differentiating g with respect to the parameters at the identity, that is, $a = 1$, $b = c = 0$, show that a basis for $T_eG = \mathfrak{sl}(2)$ is

$$\mathbf{e} = \begin{pmatrix} 0 & 1 \\ 0 & 0 \end{pmatrix}, \qquad \mathbf{f} = \begin{pmatrix} 0 & 0 \\ 1 & 0 \end{pmatrix}, \qquad \mathbf{h} = \begin{pmatrix} 1 & 0 \\ 0 & -1 \end{pmatrix}. \qquad (3.24)$$

Hint: adapt the argument of Exercise 2.2.10; see also Exercise 1.5.7. This basis is sometimes called the Chevalley basis and the names $\mathbf{h}$, $\mathbf{e}$ and $\mathbf{f}$ are standard, although there are many others in use, particularly in the physics literature. A common one is

$$J^- = \begin{pmatrix} 0 & 0 \\ 1 & 0 \end{pmatrix}, \qquad J^+ = \begin{pmatrix} 0 & 1 \\ 0 & 0 \end{pmatrix}, \qquad J^0 = \tfrac{1}{2} \begin{pmatrix} 1 & 0 \\ 0 & -1 \end{pmatrix}.$$

These are sometimes labelled as $J^- \left(\tfrac{1}{2}\right)$ and so on, to indicate that the dimension of the representation is $2 \left(\tfrac{1}{2}\right) + 1$.

Since the Lie bracket is bilinear, it suffices to give a 'multiplication table' for the basis elements. In this case show

$[\,,\,]$	$\mathbf{h}$	$\mathbf{e}$	$\mathbf{f}$
$\mathbf{h}$	0	$2\mathbf{e}$	$-2\mathbf{f}$
$\mathbf{e}$	$-2\mathbf{e}$	0	$\mathbf{h}$
$\mathbf{f}$	$2\mathbf{f}$	$-\mathbf{h}$	0

where for example the $(1, 2)$ entry means $[\mathbf{h}, \mathbf{e}] = 2\mathbf{e}$.

Exercise 3.2.18 Show that the map $\phi : \mathfrak{sl}(2) \to \mathfrak{gl}(3)$ given by

$$\phi \begin{pmatrix} a & b \\ c & -a \end{pmatrix} = \begin{pmatrix} -2a & -c & 0 \\ -2b & 0 & -2c \\ 0 & -b & 2a \end{pmatrix}$$

is a faithful representation of $\mathfrak{sl}(2)$. That is, ϕ is linear, injective, and $\phi([A, B]) = \phi(A)\phi(B) - \phi(B)\phi(A)$ for all $A, B \in \mathfrak{sl}(2)$, see Definition 3.2.10.

Exercise 3.2.19 The group $SL(3)$ is the group of 3×3 matrices with determinant equal to 1. Write the generic element as

$$g = \begin{pmatrix} a & b & c \\ d & e & f \\ h & k & \ell \end{pmatrix}. \tag{3.25}$$

Solving $\det(g) = 1$ for ℓ, back-substituting into g and then differentiating with respect to the parameters at the identity, $a = e = 1$, $b = c = d = f = h = k = 0$, yields a basis for $\mathfrak{sl}(3)$. For example, the basis elements corresponding to the parameters a and h are

$$v(a) = \frac{\partial g}{\partial a}\bigg|_{id} \begin{pmatrix} 1 & 0 & 0 \\ 0 & 0 & 0 \\ 0 & 0 & -1 \end{pmatrix}, \qquad v(h) = \begin{pmatrix} 0 & 0 & 0 \\ 0 & 0 & 0 \\ 1 & 0 & 0 \end{pmatrix}$$

respectively. Find all eight basis elements and construct the Lie bracket multiplication table. How many copies of $\mathfrak{sl}(2)$ can you find inside $\mathfrak{sl}(3)$? Hint: a less obvious one is the adjoint representation of $\mathfrak{sl}(2)$ constructed in Example 3.2.27.

Exercise 3.2.20 Find a basis and the Lie bracket multiplication table for the Lie algebra of the group $SE(2)$ of rotations and translations in the plane, by considering its standard matrix representation,

$$g(\theta, a, b) = \begin{pmatrix} \cos\theta & -\sin\theta & a \\ \sin\theta & \cos\theta & b \\ 0 & 0 & 1 \end{pmatrix}.$$

Exercise 3.2.21 Let I_2 be the 2×2 identity matrix and let J be the 4×4 matrix

$$J = \begin{pmatrix} 0 & I_2 \\ -I_2 & 0 \end{pmatrix}.$$

Define the *symplectic* group $Sp(2)$ to be the set

$$Sp(2) = \{A \in GL(4) \mid A^T J A = J\}.$$

Show that $Sp(2)$ has dimension 10 and that the Lie algebra $T_e(Sp(2)) = \mathfrak{sp}(2)$ is given by

$$\mathfrak{sp}(2) = \{X \in \mathfrak{gl}(4) \mid X^T J + J X = 0\}.$$

Hint: adapt the proof of Exercise 2.2.11. Show that a typical element of $\mathfrak{sp}(2)$ has the form

$$\begin{pmatrix} A & B \\ C & -A^T \end{pmatrix}$$

where A is an arbitrary 2×2 matrix and B and C are both symmetric, that is, $B = B^T$ and $C = C^T$. Hence write down a basis for $\mathfrak{sp}(2)$. Consider the six dimensional subgroup H of $Sp(2)$ that fixes the vector $(1\ 0\ 0\ 0)^T$ under left multiplication. Show that the Lie algebra $\mathfrak{h}$ of H sends $(1\ 0\ 0\ 0)^T$ to zero under left multiplication. Write down a basis for $\mathfrak{h}$ and construct its Lie bracket multiplication table as in Example 3.2.17. Show $\mathfrak{h} = \mathfrak{a} \otimes \mathfrak{b}$ where $\mathfrak{a} \approx \mathfrak{sl}(2)$, $\mathfrak{b}$ is a solvable Lie algebra (see Definition 4.6.7) containing an element that commutes with all of $\mathfrak{h}$ and if $x \in \mathfrak{a}$, $y \in \mathfrak{b}$ then $[x, y] \in \mathfrak{b}$.

Exercise 3.2.22 Let I_2 be the 2×2 identity matrix and let S be the 4×4 matrix

$$S = \begin{pmatrix} 0 & I_2 \\ I_2 & 0 \end{pmatrix}.$$

Define the *orthogonal* group $O(4)$ to be the set

$$O(4) = \{A \in GL(4) \mid A^T SA = S\}.$$

Show that $O(4)$ is a group of dimension six. Show that the Lie algebra $T_eO(4) = \mathfrak{o}(4)$ is given by

$$\mathfrak{o}(4) = \{X \in \mathfrak{gl}(4) \mid X^T S + SX = 0\}.$$

Show that a typical element of $\mathfrak{o}(4)$ has the form

$$\begin{pmatrix} P & \lambda K \\ \mu K & -P^T \end{pmatrix}$$

where

$$K = \begin{pmatrix} 0 & -1 \\ 1 & 0 \end{pmatrix}, \quad \lambda, \mu \in \mathbb{R}$$

and P is an arbitrary 2×2 real matrix. Hence write down a basis for $\mathfrak{o}(4)$.

Exercise 3.2.23 Exercise 3.2.22 continued. Show that $\phi_1, \phi_2 : \mathfrak{sl}(2) \to \mathfrak{o}(4)$ given by

$$\phi_1(A) = \begin{pmatrix} A & 0 \\ 0 & -A^T \end{pmatrix}, \qquad \phi_2 \begin{pmatrix} a & b \\ c & -a \end{pmatrix} = \begin{pmatrix} aI_2 & bK \\ -cK & -aI_2 \end{pmatrix}$$

are representations of $\mathfrak{sl}(2)$. Show that any element in $\mathfrak{o}(4)$ can be written in the form $\phi_1(A) + \phi_2(B)$ for some $A, B \in \mathfrak{sl}(2)$. Show that $\phi_1(\mathfrak{sl}(2)) \cap$

$\phi_2(\mathfrak{sl}(2)) = \{0\}$. Show that $[\phi_1(A), \phi_2(B)] = 0$ for all $A, B \in \mathfrak{sl}(2)$. Conclude that

$$\mathfrak{o}(4) \approx \mathfrak{sl}(2) \oplus \mathfrak{sl}(2).$$

Exercise 3.2.24 Show that $\mathbb{R}^3$ is a Lie algebra with bracket being the standard vector cross product. Construct the Lie bracket multiplication table. Show that a change of basis to make the table the same as that for $\mathfrak{sl}(2)$ requires complex coefficients. Is $\mathbb{R}^3$ isomorphic to either of $\mathfrak{sl}(2, \mathbb{C})$ or $\mathfrak{sl}(2, \mathbb{R})$? Is $\mathbb{C}^3$ with the vector cross product isomorphic to either of $\mathfrak{sl}(2, \mathbb{C})$ or $\mathfrak{sl}(2, \mathbb{R})$?

The *Ad* action, see Definition 3.2.7, is readily calculated for matrix groups. Since the product rule of differentiation holds for matrices, we have for fixed $g \in G$ that

$$Ad_g : \mathfrak{g} \to \mathfrak{g}, \qquad Ad_g(B) = g^{-1}Bg. \qquad (3.26)$$

Exercise 3.2.25 Let $G = SL(2)$. Calculate Ad_g on the standard basis of $\mathfrak{sl}(2)$ given in Exercise 3.2.17, for generic g.

Exercise 3.2.26 Show directly from Definition 3.2.11 that if $\mathfrak{g}$ is a matrix Lie algebra then

$$ad_A(B) = [A, B],$$

thus confirming Theorem 3.2.13 for matrix Lie algebras.

In the next example, a matrix representation of the adjoint action of $\mathfrak{sl}(2)$ on itself is calculated.

Example 3.2.27 Let $\mathfrak{g} = \mathfrak{sl}(2)$. Considered as a vector space, $\mathfrak{sl}(2)$ is three dimensional. So, $\mathfrak{gl}(\mathfrak{g})$ can be viewed as the set of 3×3 matrices. Recall the basis for $\mathfrak{sl}(2)$ given in Example 3.2.17. Let

$$\mathbf{h} \leftrightarrow \begin{pmatrix} 1 \\ 0 \\ 0 \end{pmatrix}, \qquad \mathbf{e} \leftrightarrow \begin{pmatrix} 0 \\ 1 \\ 0 \end{pmatrix}, \qquad \mathbf{f} \leftrightarrow \begin{pmatrix} 0 \\ 0 \\ 1 \end{pmatrix}.$$

Looking at the Lie bracket multiplication table for $\mathfrak{sl}(2)$, we need the matrix representing $ad_\mathbf{h}$ to be such that

$$ad_\mathbf{h} \begin{pmatrix} 1 \\ 0 \\ 0 \end{pmatrix} = \begin{pmatrix} 0 \\ 0 \\ 0 \end{pmatrix}, \quad ad_\mathbf{h} \begin{pmatrix} 0 \\ 1 \\ 0 \end{pmatrix} = \begin{pmatrix} 0 \\ 2 \\ 0 \end{pmatrix}, \quad ad_\mathbf{h} \begin{pmatrix} 0 \\ 0 \\ 1 \end{pmatrix} = \begin{pmatrix} 0 \\ 0 \\ -2 \end{pmatrix},$$

and thus

$$ad_{\mathbf{h}} = \begin{pmatrix} 0 & 0 & 0 \\ 0 & 2 & 0 \\ 0 & 0 & -2 \end{pmatrix}.$$

Similarly,

$$ad_{\mathbf{e}} = \begin{pmatrix} 0 & 0 & 1 \\ -2 & 0 & 0 \\ 0 & 0 & 0 \end{pmatrix}, \qquad ad_{\mathbf{f}} = \begin{pmatrix} 0 & -1 & 0 \\ 0 & 0 & 0 \\ 2 & 0 & 0 \end{pmatrix}.$$

It can be checked that

$$[ad_{\mathbf{h}}, ad_{\mathbf{e}}] = ad_{[\mathbf{h},\mathbf{e}]} = ad_{2\mathbf{e}} = 2ad_{\mathbf{e}}$$

and so forth.

Recall the definition of a faithful representation.

Definition 3.2.28 A representation $\phi : \mathfrak{g} \to \mathfrak{gl}(V)$ is said to be *faithful* if $\ker \phi = \{0\}$, that is, if $\phi(x) = 0$ implies $x = 0$.

A faithful representation can always be constructed, a result known as Ado's Theorem. The construction begins with the adjoint representation as this representation exists for every Lie algebra. Clearly ad_x is the zero map if and only if $x \in \mathfrak{g}$ commutes with every element of $\mathfrak{g}$, so the challenge in constructing a faithful representation is to include such elements, see de Graaf (2000).

3.2.2 The Lie algebra bracket for transformation groups, and Lie's Three Theorems

In Section 2.6, we argued that given a transformation group $\mathcal{T}$ acting on a manifold M, then the concept analogous to the tangent space at the identity element is the subspace $\mathcal{X}_T(M)$ of the set of all vector fields $\mathcal{X}(M)$ on M.

We briefly recall the construction of $\mathcal{X}_T(M)$. Let $\gamma(t)$ be a path based at the identity of $\mathcal{T}$; this is a set of smooth invertible maps of M parametrised by t, such that $\gamma(0) = \mathrm{id}|_M$, the identity map on M. Then $\gamma(t)$ induces a vector field X on M as follows:

$$z \mapsto X(z) = \left.\frac{\mathrm{d}}{\mathrm{d}t}\right|_{t=0} \gamma(t)(z).$$

The set of all such vector fields for all paths based at the identity in $\mathcal{T}$ is denoted by $\mathcal{X}_T(M)$. In Section 2.6 we proved that $\mathcal{X}_T(M)$ is a vector space, and that the induced flows were in $\mathcal{T}$.

The central content of Section 3.1 can be summed up as the following theorem.

Theorem 3.2.29 *The space of vector fields $\mathcal{X}(M)$ on M is a Lie algebra with the bracket being the Lie bracket defined in (3.2).*

The first theorem we prove in this section is known as Lie's Second Theorem.

Theorem 3.2.30 (Lie's Second Theorem) *If $\mathcal{T}$ is a transformation group of M, then the vector space $\mathcal{X}_T(M)$ is closed under the Lie bracket of vector fields, and hence is a Lie algebra.*

Proof If $X, Y \in \mathcal{X}_T(M)$ then $\Gamma_t^X, \Gamma_s^Y \in \mathcal{T}$ for s, t near 0. Hence

$$h(s, t) = \left(\Gamma_t^X\right)^{-1} \Gamma_s^Y \Gamma_t^X \in \mathcal{T}$$

and $h(0, t) = \mathrm{id}_M$, for all t near 0. Therefore

$$\left.\frac{\partial}{\partial s}\right|_{s=0} h = \left(D\Gamma_t^X\right)^{-1} Y(\Gamma_t^X) \in \mathcal{X}_T(M).$$

Since $\mathcal{X}_T(M)$ is a vector space over $\mathbb{R}$ and therefore is closed under limiting processes,

$$\frac{\mathrm{d}}{\mathrm{d}t} \left(D\Gamma_t^X\right)^{-1} Y(\Gamma_t^X) \in \mathcal{X}_T(M)$$

and in particular, $[X, Y] \in \mathcal{X}_T(M)$. $\qquad\square$

It is important to realise that the dimension of a Lie algebra of vector fields has nothing to do with the dimension of the space on which they are defined; this will be become clear when we look at the examples below. First of all, the vector space $\mathcal{X}(M)$ is said to be an 'infinite dimensional' Lie algebra since we allow the vector fields to have coefficients that vary with x. For example if $M = \mathbb{R}$ and we restrict to analytic vector fields, then writing the generic vector field $f(x)\partial_x$ in terms of the Taylor coefficients of f, it can be seen that a 'basis' of the analytic part of $\mathcal{X}(M)$ is given by $\{x^n \partial_x \mid n = 0, 1, 2, \ldots\}$ (this is not really a basis since, while every analytic vector field has an expression of the form $\sum a_j x^j \partial_x$, not every such sum is analytic, since convergence in an open interval on $\mathbb{R}$ is required).

Transformation groups arising as symmetries of differential equations have a finite dimensional part depending only on constants, the number of independent constants giving the dimension of the finite part of the group, and an 'infinite dimensional' part depending on functions. These functions will be solutions of a partial differential system which may be null (that is, an empty set). The infinitesimals corresponding to the finite dimensional part generate a finite dimensional Lie algebra, as in Exercise 3.2.32, while those depending on functions generate a Lie pseudogroup, see for example Olver and Pohjanpelto (2008).

Sophus Lie, in his investigation of the transformation group $\mathcal{T}$ that maps the set of solutions of a given differential equation to itself, arrived at a method that allowed him to calculate $\mathcal{X}_\mathcal{T}(M)$, the set of infinitesimal symmetry vector fields, even when the direct calculation of $\mathcal{T}$ itself was intractable, see Remark 1.6.22. His so-called second key theorem concerning $\mathcal{X}_\mathcal{T}(M)$ is stated above. The third theorem can be stated as follows.

Theorem 3.2.31 (Lie's Third Theorem) *Given a Lie algebra $\mathcal{L}$ of vector fields on M, then there is a transformation group $\mathcal{T}$ such that $\mathcal{L} = \mathcal{X}_\mathcal{T}(M)$.*

The local transformation group $\mathcal{T}$ is generated by the flows of the infinitesimal vector fields $\mathcal{X}_\mathcal{T}(M)$, see for example Ovsiannikov (1982).

The concept of a Lie algebra is now so familiar to us that it is easy to be blasé about Lie's results, even to miss the point. For the fact is that a symmetry group of a differential equation typically involves highly non-linear actions. Yet the infinitesimal vector fields of these actions not only form a vector space but any basis $\{X^i\}$ satisfies the relations

$$[X^i, X^j] = \sum_k c_{ijk} X^k$$

where the c_{ijk}, if they are not constant, depend only on those functions appearing explicitly in the X^i; *there is no other dependence on the variables as would generally be the case.* At the time, this must have seemed incredible, and if you think about it, it still is.

Exercise 3.2.32 For the $SL(2)$ action acting on $M = \mathbb{R}$,

$$\widetilde{x} = \frac{ax+b}{cx+d}, \qquad ad - bc = 1,$$

the infinitesimal symmetry vector fields are

$$\mathbf{v}_a = 2x\partial_x, \qquad \mathbf{v}_b = \partial_x, \qquad \mathbf{v}_c = -x^2\partial_x, \tag{3.27}$$

so that $\mathcal{X}_\mathcal{T}(M)$ is the three dimensional real vector space $\langle \mathbf{v}_a, \mathbf{v}_b, \mathbf{v}_c \rangle_\mathbb{R}$. Note that $\mathcal{X}_\mathcal{T}(M)$ is three dimensional even though $M = \mathbb{R}$ is one dimensional. Show the table of Lie brackets is

[,]	$\mathbf{v}_a$	$\mathbf{v}_b$	$\mathbf{v}_c$
$\mathbf{v}_a$	0	$-2\mathbf{v}_b$	$2\mathbf{v}_c$
$\mathbf{v}_b$	$2\mathbf{v}_b$	0	$-\mathbf{v}_a$
$\mathbf{v}_c$	$-2\mathbf{v}_c$	$\mathbf{v}_a$	0

Show that the change of basis

$$\mathbf{v_h} = -\mathbf{v}_a, \qquad \mathbf{v_e} = -\mathbf{v}_b, \qquad \mathbf{v_f} = -\mathbf{v}_c$$

yields a multiplication table exactly the same as the standard one for $\mathfrak{sl}(2)$ given in Exercise 3.2.17, with

$$\mathbf{v_h} \leftrightarrow \mathbf{h}, \qquad \mathbf{v_e} \leftrightarrow \mathbf{e}, \qquad \mathbf{v_f} \leftrightarrow \mathbf{f}.$$

Hence in this case, $\mathcal{X}_T(M)$ is isomorphic to $\mathfrak{sl}(2)$.

Example 3.2.33 Consider the projective action of $SL(3)$ acting on the plane as

$$\tilde{x} = \frac{ax + bu + c}{hx + ku + \ell}, \qquad \tilde{u} = \frac{dx + eu + f}{hx + ku + \ell}, \tag{3.28}$$

where

$$\det \begin{pmatrix} a & b & c \\ d & e & f \\ h & k & \ell \end{pmatrix} = 1. \tag{3.29}$$

Solving equation (3.29) for ℓ, back-substituting into equations (3.28) and then differentiating $(\tilde{x}, \tilde{u})$ with respect to the parameters at the identity, $a = e = 1$, $b = c = d = f = h = k = 0$, yields the infinitesimal vector fields to be

$$\begin{aligned}
\mathbf{v}_a &= 2x\partial_x + u\partial_u \\
\mathbf{v}_b &= u\partial_x \\
\mathbf{v}_c &= \partial_x \\
\mathbf{v}_d &= x\partial_u \\
\mathbf{v}_e &= x\partial_x + 2u\partial_u \\
\mathbf{v}_f &= \partial_u \\
\mathbf{v}_h &= -x^2\partial_x - xu\partial_u \\
\mathbf{v}_k &= -xu\partial_x - u^2\partial_u.
\end{aligned} \tag{3.30}$$

Note that the representation of the Lie algebra $\mathfrak{sl}(3)$ generated by these vector fields is eight dimensional, even though the space on which they act is two dimensional.

Exercise 3.2.34 Show that the multiplication table for the representation of $\mathfrak{sl}(3)$ represented as vectors as given in equation (3.30) above, and that for the matrix representation, Exercise 3.2.19, is exactly the same under the identification of basis elements

$$\mathbf{v}_a \leftrightarrow -v(a), \qquad \mathbf{v}_b \leftrightarrow -v(b), \qquad \ldots$$

Explain the origin of the minus sign. Hint: one is a left and one a right action.

Since $\mathcal{X}_T(M)$ is a Lie algebra, its basis satisfies the involutivity condition needed to apply Frobenius' Theorem. Hence the orbits of a transformation group action foliate the space M on which it acts. Where the vector fields are linearly independent as elements in TM, the orbits have the dimension of the Lie algebra.

Exercise 3.2.35 Consider the vector fields $X = \partial_x$, $Y = x\partial_x - y\partial_y - 2z\partial_z$ in (x, y, z)-space. Show that $[X, Y] = X$ and thus by Frobenius' Theorem, the orbits of the transformation group foliate (x, y, z)-space with two dimensional surfaces away from the origin where $Y = 0$. Show the integral surfaces, also known as leaves of the foliation, are given by $z = cy^2$ where c is a non-zero constant. Hint: y^2/z is an invariant of both flows.

3.2.2.1 Prolongations of Lie algebras of vector fields
Prolonging vector fields leads to the same Lie algebra (Olver, 1993, Theorem 2.39) In effect, the prolongation yields another representation of the original Lie algebra on an enlarged space.

Exercise 3.2.36 Prolonging the vector fields (3.27) to (x, u, u_x, u_{xx}) space yields

$$\begin{aligned}
\mathrm{pr}^{(2)}\mathbf{v}_a &= 2x\partial_x - 2u_x\partial_{u_x} - 4u_{xx}\partial_{u_{xx}} \\
\mathrm{pr}^{(2)}\mathbf{v}_b &= \partial_x \\
\mathrm{pr}^{(2)}\mathbf{v}_c &= -x^2\partial_x + 2xu_x\partial_{u_x} + (2u_x + 4xu_{xx})\partial_{u_{xx}}.
\end{aligned} \qquad (3.31)$$

Show that the Lie bracket multiplication table for the $\mathrm{pr}^{(1)}\mathbf{v}_j$ is the same as for the $\mathbf{v}_j$. Hint: the variables x, u_x and u_{xx} are viewed as independent variables for this calculation, since they are independent coordinates on the prolonged space. Show that the Lie bracket multiplication table for the $\mathrm{pr}^{(n)}\mathbf{v}_j$ is the same as for the $\mathbf{v}_j$, for all n, see Exercise 1.6.20 for the infinitesimals.

Since prolonging an action leads to involutive vector fields on ever higher dimensional spaces, there is a chance that for high enough prolongations, the orbits will foliate the space with surfaces the same dimension as the group, at least in substantial parts of the space. This is one of the conditions we will need for a moving frame, discussed in Chapter 4, to exist.

3.2.2.2 An important formula

In this section, we prove a formula needed for the proof of Theorem 4.5.5 in Chapter 4. Let coordinates near the identity e of G be given by $g = g(a_1, a_2, \ldots, a_r)$ where r is the dimension of the group. Recalling Definition 1.6.11, the infinitesimals at $z = (z_1, \ldots, z_n)$ with respect to these coordinates are

$$\zeta_j^i(z) = \frac{\partial}{\partial a_j} g \cdot z_i \bigg|_{g=e},$$

so that for any path $\gamma(t)$ in G based at the identity, we have

$$\frac{d}{dt}\bigg|_{t=0} \gamma(t) \cdot z_i = \sum_k \zeta_j^i(z) a_j'(0).$$

The question is, what can we say about

$$\frac{d}{dt} \gamma(t) \cdot z, \qquad t \neq 0?$$

As usual, we investigate by considering the $SL(2)$ action on $\mathbb{R}$,

$$\widetilde{x} = \frac{ax+b}{cx+d}, \qquad ad - bc = 1.$$

We will be looking at elements close to the identity and thus we may take $d = (1+bc)/a$ since a will be close to unity. In coordinates given by (a, b, c), the identity element is $(a, b, c) = (1, 0, 0) = e$ and we know that the infinitesimals are

$$\begin{pmatrix} \dfrac{\partial}{\partial a}\widetilde{x} \\[2mm] \dfrac{\partial}{\partial b}\widetilde{x} \\[2mm] \dfrac{\partial}{\partial c}\widetilde{x} \end{pmatrix}\bigg|_e = \begin{pmatrix} 2x \\ 1 \\ -x^2 \end{pmatrix}.$$

It can be readily checked that

$$
\begin{pmatrix}
\dfrac{\partial}{\partial a}\widetilde{x} \\[2ex]
\dfrac{\partial}{\partial b}\widetilde{x} \\[2ex]
\dfrac{\partial}{\partial c}\widetilde{x}
\end{pmatrix}
=
\begin{pmatrix}
d & -b & \dfrac{cd}{a} \\[2ex]
-c & a & -\dfrac{c^2}{a} \\[2ex]
0 & 0 & \dfrac{1}{a}
\end{pmatrix}
\begin{pmatrix}
2\widetilde{x} \\
1 \\
-\widetilde{x}^2
\end{pmatrix}
\tag{3.32}
$$

where $d = (1 + bc)/a$. Interestingly, the derivatives with respect to the group parameters at a point g of the orbit other than the identity, are a linear combination of the infinitesimal vector fields at that point, and this linear transformation depends only on g. The two questions are, is this always so, and what is the matrix of group parameters appearing in equation (3.32)? The answer to the first question is 'yes'. In this section we will prove a formula for the matrix which involves right multiplication on the group. We will need this formula in the next chapter.

Given a Lie group G, recall right multiplication is the action of G on itself given by

$$
G \times G \to G, \qquad (h, g) \mapsto R_h(g) = gh.
$$

Looking at $SL(2)$ in local coordinates near the identity, we have

$$
R_{(a,b,c)}(\alpha, \beta, \gamma) = (A, B, C) = (a\alpha + c\beta, b\alpha + d\beta, a\gamma + c\delta)
$$

where $\delta = (1 + \beta\gamma)/\alpha$. The Jacobian of $R_{(a,b,c)}$ at the point (α, β, γ) is then

$$
T_{(\alpha,\beta,\gamma)}R_{(a,b,c)} =
\begin{pmatrix}
\dfrac{\partial A}{\partial \alpha} & \dfrac{\partial A}{\partial \beta} & \dfrac{\partial A}{\partial \gamma} \\[2ex]
\dfrac{\partial B}{\partial \alpha} & \dfrac{\partial B}{\partial \beta} & \dfrac{\partial B}{\partial \gamma} \\[2ex]
\dfrac{\partial C}{\partial \alpha} & \dfrac{\partial C}{\partial \beta} & \dfrac{\partial C}{\partial \gamma}
\end{pmatrix}
$$

$$
=
\begin{pmatrix}
a & c & 0 \\[1ex]
b & d & 0 \\[1ex]
-\dfrac{c\delta}{\alpha} & \dfrac{c\gamma}{\alpha} & a + \dfrac{\beta c}{\alpha}
\end{pmatrix}.
$$

It can be checked that the matrix of group parameters appearing in equation (3.32) is given by

$$
\left(T_e R_{(a,b,c)}\right)^{-T}.
$$

In words, evaluate $T_{(\alpha,\beta,\gamma)} R_{(a,b,c)}$ at $(\alpha,\beta,\gamma) = (1,0,0)$ and take the inverse transpose.

The result we will prove is the following.

Theorem 3.2.37 *Let $(a_1, a_2, \ldots, a_r)$ be coordinates about the identity element $e = (0, 0, \ldots, 0)$ in G, and let $h \in G$ be in the coordinate chart domain. Then for each coordinate z_i of z,*

$$
\left. \begin{pmatrix} \dfrac{\partial g \cdot z_i}{\partial a_1} \\ \vdots \\ \dfrac{\partial g \cdot z_i}{\partial a_r} \end{pmatrix} \right|_{g=h} = (T_e R_h)^{-T} \begin{pmatrix} \zeta_1^i(h \cdot z) \\ \vdots \\ \zeta_r^i(h \cdot z) \end{pmatrix}. \tag{3.33}
$$

Proof Since the argument is the same for each coordinate of z, we suppress the indices on z and the corresponding index on the infinitesimal ζ. Set $R_h(a_1, \ldots, a_r) = (A_1, \ldots, A_r)$ where $A_i = A_i(a_1, \ldots, a_r)$ so that $h = (A_1, \ldots, A_r)|_e$. Then by the chain rule,

$$
\zeta_j(h \cdot z) = \left. \frac{\partial}{\partial a_j}(g \cdot h \cdot z) \right|_{g=e} = \left. \left(\sum_k \frac{\partial(g \cdot h \cdot z)}{\partial A_k} \frac{\partial A_k}{\partial a_j} \right) \right|_{g=e}
$$

so that

$$
\begin{aligned}
(\zeta_1(h \cdot z), \ldots, \zeta_r(h \cdot z)) &= \left. \left(\frac{\partial(g \cdot h \cdot z)}{\partial A_1}, \ldots, \frac{\partial(g \cdot h \cdot z)}{\partial A_r} \right) \right|_{g=e} T_e R_h \\
&= \left. \left(\frac{\partial(g \cdot z)}{\partial a_1}, \ldots, \frac{\partial(g \cdot z)}{\partial a_r} \right) \right|_{g=h} T_e R_h.
\end{aligned}
$$

Rearranging yields the result. $\qquad\square$

Another way to understand the proof is as follows. Let $g(t)$ be a path in G. Then for either a left or a right action,

$$
\left. \frac{d}{dt} \right|_{t=s} g(t) \cdot z = \left. \frac{d}{dt} \right|_{t=s} g(t)g(s)^{-1} \cdot (g(s) \cdot z) = \mathbf{v} \cdot (g(s) \cdot z) \tag{3.34}
$$

where

$$
\mathbf{v} = \left. \frac{d}{dt} \right|_{t=s} (g(t)g(s)^{-1}) \in T_e G.
$$

Now

$$\frac{d}{dt}\bigg|_{t=s}(g(t)g(s)^{-1}) = \frac{d}{dt}\bigg|_{t=s}R_{g(s)^{-1}}(g(t))$$

$$= T_{g(s)}R_{g(s)^{-1}}\frac{d}{dt}\bigg|_{t=s}g(t)$$

$$= \left(T_eR_{g(s)}\right)^{-1}g'(s),$$

using the result of Exercise 3.2.38. This second proof of Theorem 3.2.37 concludes by noting that since the infinitesimals are the coefficients, with respect to some coordinate basis of T_eG of the vector $d/dt|_{t=0}g(t) \cdot z \in T_zM$, the map induced on them will be the transpose of the linear map on T_eG with respect to that basis.

Exercise 3.2.38 Use $R_{g^{-1}}R_g = R_{gg^{-1}} = R_e$ which is the identity map, to show that

$$(T_eR_g)^{-1} = T_gR_{g^{-1}}.$$

Exercise 3.2.39 Consider the action

$$\widetilde{x} = \frac{ax+b}{cx+d}, \qquad \widetilde{u}(\widetilde{x}) = u(x), \qquad ad-bc = 1.$$

Calculate the infinitesimals $\phi_{[x],a}, \phi_{[x],c}, \phi_{[x],c}$, where

$$\phi_{[x],a} = \frac{\partial}{\partial a}\bigg|_e \widetilde{u_x}$$

and similarly for b and c, where e is the identity element given by $(a, b, c) = (1, 0, 0)$, for the induced action on $\widetilde{u_x}$, and verify

$$\begin{pmatrix} \frac{\partial}{\partial a}\widetilde{u_x} \\ \frac{\partial}{\partial b}\widetilde{u_x} \\ \frac{\partial}{\partial c}\widetilde{u_x} \end{pmatrix} = (T_eR_{(a,b,c)})^{-T}\begin{pmatrix} \phi_{[x],a}(\widetilde{x}, \widetilde{u}, \widetilde{u_x}) \\ \phi_{[x],b}(\widetilde{x}, \widetilde{u}, \widetilde{u_x}) \\ \phi_{[x],c}(\widetilde{x}, \widetilde{u}, \widetilde{u_x}) \end{pmatrix}$$

where $(T_eR_{(a,b,c)})^{-T}$ is the same matrix appearing in equation (3.32). Do the same for the induced action on $\widetilde{u_{xx}}$, thus verifying Theorem 3.2.37 for these two cases.

3.2.2.3 Lie's First Theorem

We have discussed Lie's Second and Third Theorems above. We now briefly state and prove Lie's First Theorem for completeness. Let G be a local Lie group with parameters $(a_1, \ldots)$ such that the identity element corresponds to $a_i = 0$, for all i. We note the general statement of the theorem is not restricted to groups with finitely many parameters. Let the multiplication law be given, in parameter form, as

$$\mu : ((a_1, \ldots), (b_1, \ldots)) \mapsto (\mu(a, b)_1, \ldots) = \mu(a, b).$$

Recall the definition of left multiplication,

$$L_a : G \to G, \qquad b \mapsto ab.$$

and that of its tangent map,

$$\left. \frac{\mathrm{d}}{\mathrm{d}t} \right|_{t=0} L_a(b(t)) = T_{b(0)} L_a(b'(0)).$$

The associative law, $\mu(a, \mu(b, c)) = \mu(\mu(a, b), c)$ can be written as

$$L_a(L_b(c)) = L_{\mu(a,b)}(c)$$

and the chain rule gives

$$T_b L_a \circ T_e L_b = T_e L_{\mu(a,b)}. \tag{3.35}$$

Setting

$$A(a) = T_e L_a, \qquad \psi = L_a$$

into (3.35) yields 'Lie's first equation',

$$T_b \psi = A(\psi(b)) A(b)^{-1}. \tag{3.36}$$

Note by Exercise 3.2.38, $A(a) = T_e L_a$ has an inverse. Thus, Lie's first equation is the infinitesimal form of the associative law, in some sense.

Lie's First Theorem starts with equation (3.36) and gives conditions under which a group multiplication law can be constructed.

Theorem 3.2.40 (Lie's First Theorem) *Let G be a smooth space and fix a point '0'$\in G$. Assume for all $b \in G$, there exists a linear, bijective map $A(b)$: $T_0 G \to T_b G$. If for all $a \in G$, equation (3.36) has a solution $\psi = \psi_a : G \to G$ satisfying $\psi_a(0) = a$, then G is a (local) Lie group with multiplication law $\mu(a, b) = \psi_a(b)$ and identity element 0.*

If G is described by parameters $(a_1, \ldots)$ then 0 can well be the point given by $a_i = 0$, all i. If the space G is infinite dimensional, then the tangent spaces $T_a G$ need to be interpreted in some appropriate way, and the existence and uniqueness of solutions of first order differential equations need to be proven to hold.

Proof We show the four properties needed for G to be a group.

- The property of *closure* under the multiplication law is guaranteed by the hypothesis that ψ_a maps into G.
- To show 0 *is the identity*, by hypothesis, $\mu(a, 0) = a$ for all a. To show $\mu(0, b) = b$, we show that ψ_0 is necessarily the identity map. Invoking the existence and uniqueness of solutions of first order differential equations, we note that both the identity map and ψ_0 solve the differential equation (3.36) with the same boundary condition $\psi_0(0) = 0 = \mathrm{id}|_G(0)$, and hence are the same function.
- To show the *associative law*, we note that both $\psi_a \circ \psi_b$ and $\psi_{\mu(a,b)}$ solve the differential equation (3.36) with the same value at 0,

$$\psi_a \circ \psi_b(0) = \psi_a(b) = \mu(a, b) = \psi_{\mu(a,b)}(0)$$

and hence are the same function. Thus

$$\mu(a, \mu(b, c)) = \psi_a(\psi_b(c)) = \psi_{\mu(a,b)}(c) = \mu(\mu(a, b), c).$$

- Finally, given $a \in G$ we need to exhibit an *inverse* $a^{-1} \in G$ such that $\mu(a, a^{-1}) = 0 = \mu(a^{-1}, a)$. Considering equation (3.36) and noting the assumptions on $A(a)$, we see that for any solution ψ, $T_b\psi$ has an inverse for all b and hence ψ is invertible at least locally near 0. Taking the inverse of both sides of equation (3.36), noting that $(T_b\psi)^{-1} = T_{\psi(b)}\psi^{-1}$ and setting $\beta = \psi(b)$ yields

$$T_\beta\psi^{-1} = A(\psi^{-1}(\beta))A(\beta).$$

In other words, ψ^{-1} satisfies Lie's first equation. If $\psi = \psi_a$, then for a sufficiently close to 0, we can define $a^{-1} = \psi_a^{-1}(0)$, so that $\psi_a^{-1} = \psi_{a^{-1}}$. It is straightforward to check that a^{-1} has the required properties. $\quad\square$

Exercise 3.2.41 Fill in the details of the proof of the associative law. Note that b in equation (3.36) is a dummy variable, while that in both $\psi_a \circ \psi_b$ and $\psi_{\mu(a,b)}$ is fixed.

3.3 The Adjoint and adjoint actions for transformation groups

In this section, we consider the Adjoint action Ad of a transformation group $\mathcal{T}$ on its Lie algebra $\mathcal{X}_\mathcal{T}$. We will use the formulae proved here in Chapter 7.

Astute readers will have realised that we have already used the analogues of the Ad and ad actions, given in Definitions 3.2.7 and 3.2.11, for transformation groups, in the proof of Theorem 3.2.30.

Definition 3.3.1 Given a transformation group $\mathcal{T}$ acting on M, the induced Adjoint action Ad of G on $\mathcal{X}(M)$ is

$$(g, X) \mapsto Ad_g(X), \qquad Ad_g(X)(x) = Tg^{-1}X(g \cdot x)$$

where $Tg : TM \to TM$ is the tangent map of g considered as a map $g : M \to M$.

Recalling that a vector field is a map $X : M \to TM$ such that the base point of $X(x)$ is x, we have that the diagram

$$
\begin{array}{ccc}
TM & \overset{Tg}{\to} & TM \\
Ad_g(X) \uparrow & & \uparrow X \\
M & \overset{g}{\to} & M
\end{array}
\qquad (3.37)
$$

commutes.

Thus, for a vector field given in column vector notation, we have given coordinates $z = (z_1, z_2, \ldots, z_n)$ and with $g \cdot z = \widetilde{z} = (\widetilde{z}_1, \widetilde{z}_2, \ldots, \widetilde{z}_n)$,

$$
Ad_g(\mathbf{f})(z) =
\begin{pmatrix}
\dfrac{\partial \widetilde{z}_1}{\partial z_1} & \cdots & \dfrac{\partial \widetilde{z}_1}{\partial z_n} \\
\vdots & \ddots & \vdots \\
\dfrac{\partial \widetilde{z}_n}{\partial z_1} & \cdots & \dfrac{\partial \widetilde{z}_n}{\partial z_n}
\end{pmatrix}^{-1}
\begin{pmatrix}
f_1(\widetilde{z}) \\
\vdots \\
f_n(\widetilde{z})
\end{pmatrix}
= \frac{\partial \widetilde{z}}{\partial z}^{-1} \mathbf{f}(\widetilde{z})
\qquad (3.38)
$$

where the last line defines the matrix $\partial \widetilde{z}/\partial z$, the Jacobian of the map $z \mapsto g \cdot z = \widetilde{z}$, which is of course, the map Tg in coordinates.

Exercise 3.3.2 If $X = f(x)\partial_x$ and $g \cdot x = \widetilde{x}$, show $Ad_g(X) = f(\widetilde{x})\partial_{\widetilde{x}}$.

For a vector written in operator notation, the Ad action of G on a vector field is, given coordinates $z = (z_1, z_2, \ldots, z_n)$ and with $g \cdot z = \widetilde{z} = (\widetilde{z}_1, \widetilde{z}_2, \ldots, \widetilde{z}_n)$,

$$
Ad_g \left(\sum_i f_i(z) \frac{\partial}{\partial z_i} \right) = \sum f_i(\widetilde{z}) \frac{\partial}{\partial \widetilde{z}_i}.
\qquad (3.39)
$$

Indeed,

$$
Ad_g \left(\sum_i f_i(z) \frac{\partial}{\partial z_i} \right)
$$

$$
= (f_1(\widetilde{z}), \ldots, f_n(\widetilde{z}))
\begin{pmatrix}
\dfrac{\partial \widetilde{z}_1}{\partial z_1} & \cdots & \dfrac{\partial \widetilde{z}_n}{\partial z_1} \\
\vdots & \ddots & \vdots \\
\dfrac{\partial \widetilde{z}_1}{\partial z_n} & \cdots & \dfrac{\partial \widetilde{z}_n}{\partial z_n}
\end{pmatrix}^{-1}
\begin{pmatrix}
\dfrac{\partial}{\partial z_1} \\
\vdots \\
\dfrac{\partial}{\partial z_n}
\end{pmatrix}
$$

$$= (f_1(\widetilde{z}), \dots, f_n(\widetilde{z})) \begin{pmatrix} \dfrac{\partial z_1}{\partial \widetilde{z}_1} & \cdots & \dfrac{\partial z_n}{\partial \widetilde{z}_1} \\ \vdots & \ddots & \vdots \\ \dfrac{\partial z_1}{\partial \widetilde{z}_n} & \cdots & \dfrac{\partial z_n}{\partial \widetilde{z}_n} \end{pmatrix} \begin{pmatrix} \dfrac{\partial}{\partial z_1} \\ \vdots \\ \dfrac{\partial}{\partial z_n} \end{pmatrix}$$

$$= \sum f_i(\widetilde{z}) \frac{\partial}{\partial \widetilde{z}_i}.$$

Of particular interest is the Ad action restricted to the infinitesimal vector fields. We first verify that Ad_g is an action on $\mathcal{X}_T(M)$ as given in Definition 3.2.7, and prove some of Exercise 3.2.9 for the particular case of transformation groups.

Lemma 3.3.3 *Let T be a transformation group, and let $\mathbf{v} \in \mathcal{X}_T(M)$ be an infinitesimal vector field. For $g \in T$,*

$$Ad_g(\mathbf{v}) \in \mathcal{X}_T(M).$$

Proof Let $\Gamma_t^{\mathbf{v}}$ be the flow on M induced by $\mathbf{v}$. Then $g^{-1}\Gamma_t^{\mathbf{v}}g$ is a path in T, and, at $t = 0$, is the identity map. Hence

$$Ad_g(\mathbf{v}) = Tg^{-1}\mathbf{v}g = \frac{\mathrm{d}}{\mathrm{d}t}\Big|_{t=0} g^{-1}\Gamma_t^{\mathbf{v}}g \in \mathcal{X}_T(M). \qquad \square$$

Lemma 3.3.4 *For $g, h \in T$, $\mathbf{v} \in \mathcal{X}_T(M)$, and assuming a left action of T on M,*

$$Ad_h\left(Ad_g(\mathbf{v})\right) = Ad_{gh}(\mathbf{v}).$$

Proof Using diagram (3.37) twice, we have

$$\begin{array}{ccccc} & TM & \overset{Th}{\to} & TM & \overset{Tg}{\to} & TM \\ Ad_h(Ad_g(X))\uparrow & & Ad_g(X)\uparrow & & \uparrow X \\ & M & \overset{h}{\to} & M & \overset{g}{\to} & M \end{array}$$

and

$$\begin{aligned} Ad_h\left(Ad_g(\mathbf{v})\right)(z) &= Th^{-1}\big|_{h\cdot z}\left(Ad_g\mathbf{v}\right)(h\cdot z) \\ &= Th^{-1}\big|_{h\cdot z}Tg^{-1}\big|_{g\cdot(h\cdot z)}\mathbf{v}(g\cdot(h\cdot z)) \\ &= (Tg \circ Th)^{-1}\big|_{(gh)\cdot z}\mathbf{v}((gh)\cdot z) \\ &= T(gh)^{-1}\big|_{(gh)\cdot z}\mathbf{v}((gh)\cdot z) \\ &= Ad_{gh}(\mathbf{v})(z). \end{aligned} \qquad \square$$

Since Tg is a linear map, then Ad_g is linear on $\mathcal{X}_T(M)$. Lemma 3.3.3 implies that for any basis $\mathbf{v}^i$, $i = 1, \ldots, r$ for $\mathcal{X}_T(M)$, where $r = \dim T$,

$$Ad_g\left(\sum_i \alpha_i \mathbf{v}^i\right) = \sum_i \alpha_i Ad_g(\mathbf{v}^i) = \sum_{i,j} \alpha_i Ad(g)^i_j \mathbf{v}^j \qquad (3.40)$$

for some $r \times r$ matrix $Ad(g)$. We next investigate the properties of $Ad(g)$ and how to compute it.

Lemma 3.3.5 *The map $g \mapsto Ad(g) \in GL(r)$, defined in equation (3.40), is a representation of T.*

Proof Let $g, h \in G$.

$$\begin{aligned}
\sum_{i,k} \alpha_i Ad(gh)^i_k \mathbf{v}^k &= \sum_i \alpha_i Ad_{gh}\mathbf{v}_i \\
&= \sum_i \alpha_i Ad_h Ad_g \mathbf{v}^i \\
&= \sum_{i,j} \alpha_i Ad(g)^i_j Ad_h \mathbf{v}^j \\
&= \sum_{i,j,k} \alpha_i Ad(g)^i_j Ad(h)^j_k \mathbf{v}^k \\
&= \sum_{i,k} \alpha_i (Ad(g)Ad(h))^i_k \mathbf{v}^k
\end{aligned}$$

where we have used the linearity of Ad_h in the third line. $\square$

If we write

$$\sum_{i,j} \alpha_i Ad(g)^i_j \mathbf{v}^j = \sum_i \widetilde{\alpha}_i \mathbf{v}^i$$

then writing $\boldsymbol{\alpha} = (\alpha_1, \ldots, \alpha_r)^T$ as a column vector, we have

$$\widetilde{\boldsymbol{\alpha}} = Ad(g)^T \boldsymbol{\alpha}. \qquad (3.41)$$

To calculate $Ad(g)$, it can be easier to obtain the action on the α_i using the form of the Adjoint action given in equation (3.39).

Example 3.3.6 For the $SL(2)$ action

$$\widetilde{x} = \frac{ax+b}{cx+d}, \qquad ad - bc = 1$$

with infinitesimal vector fields

$$\mathbf{v} = (\alpha + \beta x + \gamma x^2)\partial_x,$$

we obtain

$$
\begin{aligned}
Ad_g(\mathbf{v})(x) &= \left(\alpha + \beta\widetilde{x} + \gamma\widetilde{x}^2\right)\frac{\partial}{\partial\widetilde{x}} \\
&= \left(\alpha + \beta\left(\frac{ax+b}{cx+d}\right) + \gamma\left(\frac{ax+b}{cx+d}\right)^2\right)\left(\frac{\partial\widetilde{x}}{\partial x}\right)^{-1}\frac{\partial}{\partial x} \\
&= \left(\alpha(cx+d)^2 + \beta(ax+b)(cx+d) + \gamma(ax+b)^2\right)\frac{\partial}{\partial x} \\
&= \left(\widetilde{\alpha} + \widetilde{\beta}x + \widetilde{\gamma}x^2\right)\frac{\partial}{\partial x}
\end{aligned}
$$

so that

$$
\begin{pmatrix} \widetilde{\alpha} \\ \widetilde{\beta} \\ \widetilde{\gamma} \end{pmatrix} = Ad(g)^T \begin{pmatrix} \alpha \\ \beta \\ \gamma \end{pmatrix} = \begin{pmatrix} d^2 & bd & b^2 \\ 2cd & ad+bc & 2ab \\ c^2 & ac & a^2 \end{pmatrix} \begin{pmatrix} \alpha \\ \beta \\ \gamma \end{pmatrix}. \tag{3.42}
$$

Exercise 3.3.7 Show that for the $SL(2)$ action,

$$
\widetilde{x} = \frac{ax+b}{cx+d}, \qquad \widetilde{u} = 6c(cx+d) + (cx+d)^2 u, \qquad ad - bc = 1
$$

that $Ad(g)$ is the same as that given in equation (3.42) for the action in Example 3.3.6. Explain.

Exercise 3.3.8 For a matrix Lie group $G \subset GL(n)$ acting linearly on $\mathbb{R}^n$, the infinitesimal vector fields are the constant vector fields,

$$
V_a(\mathbf{x}) = a\mathbf{x}, \qquad \mathbf{x} \in \mathbb{R}^n,\ a \in \mathfrak{g}.
$$

Show the Adjoint action is

$$
Ad_g(V_a)(\mathbf{x}) = g^{-1}ag\mathbf{x}.
$$

What is $Ad(g)$ in the case $G = Sp(2)$?

Exercise 3.3.9 For the action of the Euclidean group acting on $\mathbb{R}^2$,

$$
\begin{pmatrix} \widetilde{x} \\ \widetilde{u} \end{pmatrix} = \begin{pmatrix} \cos\theta & -\sin\theta \\ \sin\theta & \cos\theta \end{pmatrix} \begin{pmatrix} x \\ u \end{pmatrix} + \begin{pmatrix} a \\ b \end{pmatrix},
$$

the infinitesimal vector fields are

$$
\mathbf{v} = \alpha(-u\partial_x + x\partial_u) + \beta\partial_x + \gamma\partial_u \tag{3.43}
$$

for arbitrary constants α, β, γ. Show that

$$
\begin{pmatrix} \tilde{\beta} \\ \tilde{\gamma} \\ \tilde{\alpha} \end{pmatrix} = Ad(g(\theta, a, b))^T \begin{pmatrix} \beta \\ \gamma \\ \alpha \end{pmatrix}
$$
$$
= \begin{pmatrix} \cos\theta & \sin\theta & a\sin\theta - b\cos\theta \\ -\sin\theta & \cos\theta & b\sin\theta + a\cos\theta \\ 0 & 0 & 1 \end{pmatrix} \begin{pmatrix} \beta \\ \gamma \\ \alpha \end{pmatrix}.
$$

Obtain the Adjoint representation another way by adapting the method of Exercise 3.3.8; consider the standard representation of the Euclidean group,

$$
R(g(\theta, a, b)) = \begin{pmatrix} \cos\theta & -\sin\theta & a \\ \sin\theta & \cos\theta & b \\ 0 & 0 & 1 \end{pmatrix} \in GL(3)
$$

with Lie algebra

$$
\left\{ \begin{pmatrix} 0 & -\alpha & \beta \\ \alpha & 0 & \gamma \\ 0 & 0 & 0 \end{pmatrix} \mid \alpha, \beta, \gamma \in \mathbb{R} \right\} \tag{3.44}
$$

and a linear action on $\mathbb{R}^3$ restricted to the plane $\{(x, y, 1) \mid x, y \in \mathbb{R}\}$. Show that both methods yield the same $Ad(g)$. Hint: the use of the same names for the arbitrary constants in (3.43) and (3.44) is deliberate; it indicates the isomorphism between the two presentations of the Lie algebra.

We come now to the main computational result we will need in the sequel, concerning the Ad action as induced on $\mathcal{X}_T(M)$, which is essentially equation (3.40) in matrix form.

Theorem 3.3.10 *Let coordinates on M be $z = (z_1, z_2, \ldots, z_n)$, and $g \cdot z = \tilde{z} = (\tilde{z}_1, \tilde{z}_2, \ldots, \tilde{z}_n)$. Denote the Jacobian matrix of the group action as*

$$
\frac{\partial \tilde{z}}{\partial z} = \begin{pmatrix} \dfrac{\partial \tilde{z}_1}{\partial z_1} & \dfrac{\partial \tilde{z}_1}{\partial z_2} & \cdots & \dfrac{\partial \tilde{z}_1}{\partial z_n} \\[2ex] \dfrac{\partial \tilde{z}_2}{\partial z_1} & \dfrac{\partial \tilde{z}_2}{\partial z_2} & \cdots & \dfrac{\partial \tilde{z}_2}{\partial z_n} \\[2ex] \vdots & \vdots & \ddots & \vdots \\[2ex] \dfrac{\partial \tilde{z}_n}{\partial z_1} & \dfrac{\partial \tilde{z}_n}{\partial z_2} & \cdots & \dfrac{\partial \tilde{z}_n}{\partial z_n} \end{pmatrix}. \tag{3.45}
$$

Let $(a_1, a_2, \ldots, a_r)$ be coordinates on G, or the independent parameters for T, and let the infinitesimal vector field with respect to the coordinate a_i be given as

$$\mathbf{v}^i(z) = \zeta_1^i(z)\frac{\partial}{\partial z_1} + \cdots + \zeta_n^i(z)\frac{\partial}{\partial z_n}.$$

Let the matrix $\Omega(z)$ be given by

$$\Omega(z) = \begin{array}{c} \\ a_1 \\ a_2 \\ \vdots \\ a_r \end{array} \begin{array}{cccc} z_1 & z_2 & \cdots & z_n \\ \left(\begin{array}{cccc} \zeta_1^1(z) & \zeta_2^1(z) & \cdots & \zeta_n^1(z) \\ \zeta_1^2(z) & \zeta_2^2(z) & \cdots & \zeta_n^2(z) \\ \vdots & \vdots & \ddots & \vdots \\ \zeta_1^r(z) & \zeta_2^r(z) & \cdots & \zeta_n^r(z) \end{array}\right). \end{array}$$

Let $Ad(g)$ be the $r \times r$ matrix giving the Ad action on $\mathcal{X}_T(M)$ with respect to the basis $\langle \mathbf{v}^1, \mathbf{v}^2, \ldots, \mathbf{v}^r \rangle$ of $\mathcal{X}_T(M)$ used to write down Ω, that is, $Ad_g(\mathbf{v}^i) = \sum_j Ad(g)^i_j \mathbf{v}^j$. Then

$$Ad(g)\Omega(z) = \Omega(\widetilde{z})\left(\frac{\partial \widetilde{z}}{\partial z}\right)^{-T}. \tag{3.46}$$

Proof Multiplying both sides of equation (3.46) on the right by

$$\nabla^T = \left(\frac{\partial}{\partial z_1}, \frac{\partial}{\partial z_2}, \ldots, \frac{\partial}{\partial z_n}\right)^T,$$

the ith component of the right hand side is the definition of $Ad_g(\mathbf{v}^i)$, while the ith component of the left hand side is, by construction, $Ad(g)^i_j \mathbf{v}^j$. These are equal by the definition of $Ad(g)$. $\qquad \square$

Example 3.3.11 For the action

$$\widetilde{x} = x, \qquad \widetilde{u} = \frac{au+b}{cu+d}, \qquad ad - bc = 1,$$

consider the prolonged action on (u, u_x, u_{xx}) space. The prolonged action is

$$\widetilde{u_x} = \frac{u_x}{(cu+d)^2}, \qquad \widetilde{u_{xx}} = \frac{u_{xx}}{(cu+d)^2} - 2\frac{cu_x^2}{(cu+d)^3}.$$

The infinitesimal vector fields corresponding to the parameters (a, b, c) are

$$\mathbf{v}_a = 2u\frac{\partial}{\partial u} + 2u_x\frac{\partial}{\partial u_x} + 2u_{xx}\frac{\partial}{\partial u_{xx}}$$

$$\mathbf{v}_b = \frac{\partial}{\partial u}$$

$$\mathbf{v}_c = -u^2\frac{\partial}{\partial u} - 2uu_x\frac{\partial}{\partial u_x} - 2(u_x^2 + uu_{xx})\frac{\partial}{\partial u_{xx}}$$

so that

$$\Omega(u, u_x, u_{xx}) = \begin{array}{c} \\ a \\ b \\ c \end{array}\overset{\begin{array}{ccc} u & u_x & u_{xx} \end{array}}{\left(\begin{array}{ccc} 2u & 2u_x & 2u_{xx} \\ 1 & 0 & 0 \\ -u^2 & -2uu_x & -2(u_x^2 + uu_{xx}) \end{array}\right)}$$

Next we have

$$\frac{\partial(\widetilde{u}, \widetilde{u}_x, \widetilde{u}_{xx})}{\partial(u, u_x, u_{xx})} = \left(\begin{array}{ccc} \dfrac{1}{(cu+d)^2} & 0 & 0 \\[3mm] \dfrac{-2cu_x}{(cu+d)^3} & \dfrac{1}{(cu+d)^2} & 0 \\[3mm] \dfrac{-2c((cu+d)u_{xx} - 3u_x^2)}{(cu+d)^4} & \dfrac{-4cu_x}{(cu+d)^3} & \dfrac{1}{(cu+d)^2} \end{array}\right)$$

The matrix $\mathcal{Ad}(g)$ is more or less that given in equation (3.42), adjusted to the basis we are using here,

$$\mathcal{Ad}(g) = \begin{array}{c} \\ a \\ b \\ c \end{array}\overset{\begin{array}{ccc} a & b & c \end{array}}{\left(\begin{array}{ccc} ad+bc & 2bd & -2ca \\ cd & d^2 & -c^2 \\ -ab & -b^2 & a^2 \end{array}\right)}.$$

It is straightforward to check that

$$\mathcal{Ad}(g)\Omega(u, u_x, u_{xx}) = \Omega(\widetilde{u}, \widetilde{u}_x, \widetilde{u}_{xx})\left(\frac{\partial(\widetilde{u}, \widetilde{u}_x, \widetilde{u}_{xx})}{\partial(u, u_x, u_{xx})}\right)^{-T}.$$

Exercise 3.3.12 Prolong to $(u, u_x, u_{xx}, u_{xxx}, u_{xxxx})$ space the action of Example 3.3.11. Check that

$$\mathcal{Ad}(g)\Omega(u, \ldots, u_{xxxx}) = \Omega(\widetilde{u}, \ldots, \widetilde{u}_{xxxx})\left(\frac{\partial(\widetilde{u}, \ldots, \widetilde{u}_{xxxx})}{\partial(u, \ldots, u_{xxxx})}\right)^{-T}.$$

Finally, we recall that the induced adjoint action of $\mathcal{X}_T(M)$ on $\mathcal{X}(M)$ is, for $\mathbf{v} \in \mathcal{X}_T(M)$ with flow $g(t) = \Gamma_t^{\mathbf{v}}$ and $\mathbf{w} \in \mathcal{X}_T(M)$, given by

$$ad_{\mathbf{v}}(\mathbf{w}) = \frac{d}{dt}\Big|_{t=0} Ad_{g(t)}(\mathbf{w}) = [\mathbf{v}, \mathbf{w}], \tag{3.47}$$

the standard Lie bracket of vector fields.

Exercise 3.3.13 Prove equation (3.47). Hint: compare the formulae for the definitions of the Adjoint action and the Lie bracket of vector fields.

The adjoint action ad clearly restricts to an action of $\mathcal{X}_T(M)$ to itself, and is the same as what was defined earlier as the adjoint action of any Lie algebra on itself.

4

Moving frames

In this chapter we can begin our study of the invariant calculus. The concept from which all else derives is that of a *moving frame*. We use the definition and construction as detailed by Fels and Olver (1998, 1999). Although the term 'moving frame', or 'repère mobile' is associated with Élie Cartan (1953), the idea was used, albeit implicitly, long before. A pre-Cartan history of the subject is given by Akivis and Rosenfeld (1993), and the Fels and Olver papers have a more recent historical overview. The definition of a moving frame used here has the major advantage that it can be applied to both smooth and discrete problems. In particular, there is no need for any of the paraphernalia of Differential Geometry such as exterior calculus, frame bundles and connections.[†]

4.1 Moving frames

The original problem solved by moving frames was the *equivalence problem*, 'when can two surfaces be mapped one to the other, under a coordinate trans-formation of a particular type?' It turns out there are many problems which can be formulated this way. One is the classification problem of differential equations. If you have a differential equation to solve and a database of solved equations, it is only sensible to ask, is there a coordinate transformation that takes my equation to one of the solved ones? Viewing differential equations as surfaces in $(x, u, u_x, u_{xx}, \dots)$ space, you might then apply moving frame theory. The computational complexities might be considerable, but at least you have an idea of how you might proceed.

Computer vision experts study a discrete version of the equivalence problem: given a digital image, can you match it to one of a database? Of course, the image you have and the image in the database will not be exact matches, there

---

[†] Unless, of course, Differential Geometry is your application.

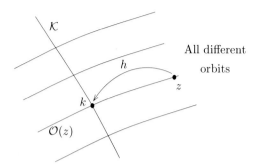

Figure 4.1 A local foliation with a transverse cross section.

may be a rotation involved or some kind of distortion, but since they are pictures of the same object, you want the computer to recognise they are equivalent up to some well defined 'distortion group'.

In later chapters, we will see other uses of moving frames. The applications all involve the fact that a moving frame defines local coordinates that provide a 'divide and conquer' mechanism for studying the problem at hand.

Moving frames exist when and where the group action is free and regular, see Definition 1.4.8. Under quite general conditions, there are ways and means of getting the group action you have to be free and regular on an extension of your space if it is not already, so the condition is not as restricting as it sounds.

If the action is free and regular in some domain Ω, then the following will be true; see Figure 4.1: For every $x \in \Omega$, there is a neighbourhood U of x such that the following hold.

- The group orbits all have the dimension of the group and foliate U.
- There is a surface $\mathcal{K} \subset U$ which crosses the group orbits transversally (see Definition 1.4.9), and for which the intersection of a given group orbit with $\mathcal{K}$ is a single point. The surface $\mathcal{K}$ is called the *cross section*.
- If $\mathcal{O}(z)$ denotes the group orbit through z, then the group element h taking $z \in U$ to $\{k\} = \mathcal{O}(z) \cap \mathcal{K}$ is unique.

The cross section $\mathcal{K}$ will not be unique. It is usually chosen to make the calculations as simple as possible.

Definition 4.1.1 The map $\rho : U \to G$ which takes a point $z \in U$ to the unique group element $\rho(z) \in G$ such that

$$\rho(z) \cdot z = k, \qquad \{k\} = \mathcal{O}(z) \cap \mathcal{K}, \qquad (4.1)$$

is called the *right moving frame* relative to the cross section $\mathcal{K}$.

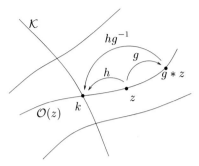

Figure 4.2 Construction of a right moving frame using a cross section.

The element h in Figure 4.1 is $\rho(z)$. If the action is left, then the map ρ is right equivariant, that is,

$$\rho(g * z) = \rho(z)g^{-1}.$$

Indeed, looking at Figure 4.2, we see that the group element taking $g * z$ to k has to be hg^{-1}. If the action is right, then the map ρ satisfies

$$\rho(g \bullet z) = g^{-1}\rho(z).$$

The inverse of the right moving frame is called the left moving frame.

More generally, we have the following definition.

Definition 4.1.2 Given a group action $G \times M \to M$, a *moving frame* is an equivariant map $\rho : M \to G$.

We have the following table.

	left action	right action
right frame	$\rho(g * z) = \rho(z)g^{-1}$	$\rho(g \bullet z) = g^{-1}\rho(z)$
left frame	$\rho(g * z) = g\rho(z)$	$\rho(g \bullet z) = \rho(z)g$

The conditions on the group action for the above construction to be valid typically hold only locally, and thus a moving frame can be constructed only locally. In general, it will not be possible to construct a moving frame in a neighbourhood of every point on a manifold: no moving frame can be constructed in the neighbourhood of the origin of $\mathbb{R}^2$ for the rotation group $SO(2)$.

We summarise the discussion as a theorem.

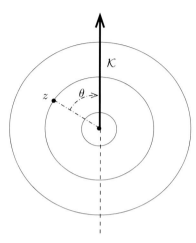

Figure 4.3 The construction of a moving frame for rotations in the plane.

Theorem 4.1.3 *If a group action is free and regular in $\Omega \subset M$, then for every $x \in \Omega$ there is a neighbourhood U of x such that there exists a moving frame on U.*

In practice, we do not wish to study the details of the foliation of the space by the group orbits. We need a way to derive the frame knowing only the formulae for the group action. Thus, we use the following procedure. We define the cross section $\mathcal{K}$ as the locus of a set of equations $\psi_k(z) = 0$, $k = 1, \ldots, r$. The number of equations, r, equals the dimension of the group. In order to obtain the group element that takes z to $\mathcal{K}$, we solve the so called *normalisation equations*,

$$\psi_k(\widetilde{z}) = 0, \qquad k = 1, \ldots, r. \tag{4.2}$$

The frame $\rho(z)$ therefore satisfies

$$\psi_i(\rho(z) * z) = 0, \qquad i = 1, \ldots, r.$$

If the solution is unique on the domain U, then ρ is a right frame, one that satisfies $\rho(g * z) = \rho(z)g^{-1}$ or $\rho(g \bullet z) = g^{-1}\rho(z)$. One chooses the normalisation equations to minimise the computations as much as possible for the application at hand. A fuller discussion appears in Section 4.2.

Example 4.1.4 Consider the usual action of $SO(2)$ in the plane,

$$\begin{pmatrix} \widetilde{x} \\ \widetilde{y} \end{pmatrix} = R_\theta \begin{pmatrix} x \\ y \end{pmatrix} = \begin{pmatrix} \cos\theta & -\sin\theta \\ \sin\theta & \cos\theta \end{pmatrix} \begin{pmatrix} x \\ y \end{pmatrix},$$

refer to Figure 4.3. If we set $U = \mathbb{R}^2 \setminus \{(0, y) \mid y \leq 0\}$, that is, we remove the origin and the negative y-axis from the plane, then the action is free and regular.

Taking the normalisation equation to be $\tilde{x} = 0$, then $\rho(x, y)$ is the rotation that takes the point (x, y) to the positive y-axis, which is the cross section $\mathcal{K}$ to the orbits. The generic element $z \in U$ has coordinates (x, y). Specifically, we have

$$\rho(x, y) = \begin{pmatrix} \dfrac{y}{\sqrt{x^2 + y^2}} & -\dfrac{x}{\sqrt{x^2 + y^2}} \\ \dfrac{x}{\sqrt{x^2 + y^2}} & \dfrac{y}{\sqrt{x^2 + y^2}} \end{pmatrix}.$$

To verify that ρ is equivariant, let R_t be a rotation matrix with angle t. Then

$$\rho(R_t * (x, y)) = \begin{pmatrix} \dfrac{\sin(t)x + \cos(t)y}{\sqrt{x^2 + y^2}} & -\dfrac{\cos(t)x - \sin(t)y}{\sqrt{x^2 + y^2}} \\ \dfrac{\cos(t)x - \sin(t)y}{\sqrt{x^2 + y^2}} & \dfrac{\sin(t)x + \cos(t)y}{\sqrt{x^2 + y^2}} \end{pmatrix}$$

$$= \begin{pmatrix} \dfrac{y}{\sqrt{x^2 + y^2}} & -\dfrac{x}{\sqrt{x^2 + y^2}} \\ \dfrac{x}{\sqrt{x^2 + y^2}} & \dfrac{y}{\sqrt{x^2 + y^2}} \end{pmatrix} \cdot \begin{pmatrix} \cos t & \sin t \\ -\sin t & \cos t \end{pmatrix}$$

$$= \rho(x, y) R_t^{-1}$$

as required. If we solve the normalisation equation for the group parameter θ, namely, $\tilde{x} = \cos(\theta)x - \sin(\theta)y = 0$, we obtain ρ not in matrix form, but parametric form,

$$\rho(x, y) = \begin{cases} \arctan\left(\dfrac{x}{y}\right) & y \neq 0 \\ 0 & y = 0 \end{cases}$$

with the usual caveats on the definition of the *arctan* function to make ρ continuous on U. To verify that ρ in this formulation is equivariant, note that

$$\rho(R_t * (x, y)) = \rho(\cos(t)x - \sin(t)y, \sin(t)x + \cos(t)y)$$

$$= \arctan\left(\frac{\cos(t)x - \sin(t)y}{\sin(t)x + \cos(t)y}\right)$$

$$= \arctan\left(\frac{(x/y) - \tan t}{1 + (x/y)\tan t}\right)$$

$$= \arctan\left(\frac{x}{y}\right) - t$$

using the addition formula for tan. Since $R_t R_s = R_{t+s}$, we see that on the parameter level, group multiplication is addition and thus the inverse of an element's parameter is its negative.

Exercise 4.1.5 Redo the calculations for Exercise 4.1.4 above but using the cross section $\mathcal{K} = \{(x, 0) \mid x > 0\}$, so that the normalisation equation is $\widetilde{y} = 0$.

Example 4.1.6 Consider now the special Euclidean group $SE(2)$ of rotations and translations acting on curves in the plane, and the induced action on its tangent lines, depicted in Figure 1.1. The induced action on derivatives is called the prolonged action detailed in Section 1.3.4. If the curve $\gamma(s)$ has coordinates $(u(s), v(s))$, the prolonged action takes place in $(u, v, u_s, v_s, u_{ss}, \dots)$-space; as many derivatives as make sense for the curve you have. The curve must be at least once differentiable, and then the prolonged group action is free on an open domain. The generic group element has parameters $(\theta, (a, b))$ where θ is the angle of rotation and (a, b) is the vector of translation. The calculations are much easier if the 'inverse' action is taken, namely

$$\begin{pmatrix} \widetilde{u} \\ \widetilde{v} \end{pmatrix} = \begin{pmatrix} \cos\theta & \sin\theta \\ -\sin\theta & \cos\theta \end{pmatrix} \begin{pmatrix} u - a \\ v - b \end{pmatrix}.$$

This is now a right action. Since the curve parameter s is invariant, the prolongation action is simple to calculate. We obtain

$$\begin{pmatrix} \widetilde{u_J} \\ \widetilde{v_J} \end{pmatrix} = \begin{pmatrix} \cos\theta & \sin\theta \\ -\sin\theta & \cos\theta \end{pmatrix} \begin{pmatrix} u_J \\ v_J \end{pmatrix}$$

where J is the index of differentiation. Let us take the cross section $\mathcal{K}$ to be the coordinate plane, $u = 0$, $v = 0$ and $v_s = 0$. Thus the normalisation equations are

$$\widetilde{u} = 0, \qquad \widetilde{v} = 0, \qquad \widetilde{v_s} = 0. \tag{4.3}$$

The action of the frame defined by these equations on a curve is depicted in Figure 4.4. Solving equations (4.3) for the group parameters in terms of $(u, v, u_s, \dots)$ yields

$$a = u, \qquad b = v, \qquad \theta = \arctan\left(\frac{v_s}{u_s}\right). \tag{4.4}$$

This is the frame in parameter form, valid on an open domain that excludes, say, the line $\{(u_s, 0) \mid u_s \leq 0\}$ in the (u_s, v_s)-plane. In matrix form, the frame is obtained by substituting the values of the parameters on the frame into a matrix representation of the generic group element. For the standard representation of

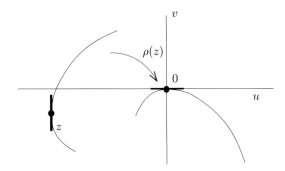

Figure 4.4 The action of the standard Euclidean frame group element, defined by $\widetilde{u} = 0$, $\widetilde{v} = 0$ and $\widetilde{v}_s = 0$, at the point $z = (u, v, u_s, v_s)$.

$SE(2)$ (see equation (1.21)) we obtain

$$\rho(u, v, u_s, v_s, \ldots) = \begin{pmatrix} \dfrac{u_s}{\sqrt{u_s^2 + v_s^2}} & -\dfrac{v_s}{\sqrt{u_s^2 + v_s^2}} & u \\[3mm] \dfrac{v_s}{\sqrt{u_s^2 + v_s^2}} & \dfrac{u_s}{\sqrt{u_s^2 + v_s^2}} & v \\[3mm] 0 & 0 & 1 \end{pmatrix}. \tag{4.5}$$

To verify that this is equivariant, we act on the components of ρ with a generic group element which we take to have parameters $(\alpha, (k_1, k_2))$; it is not a good idea to take the same parameter names as those used to calculate the frame. Then

$$\rho(\widetilde{u}, \widetilde{v}, \widetilde{u}_s, \ldots) = \begin{pmatrix} \dfrac{\widetilde{u}_s}{\sqrt{\widetilde{u}_s^2 + \widetilde{v}_s^2}} & -\dfrac{\widetilde{v}_s}{\sqrt{\widetilde{u}_s^2 + \widetilde{v}_s^2}} & \widetilde{u} \\[3mm] \dfrac{\widetilde{v}_s}{\sqrt{\widetilde{u}_s^2 + \widetilde{v}_s^2}} & \dfrac{\widetilde{u}_s}{\sqrt{\widetilde{u}_s^2 + \widetilde{v}_s^2}} & \widetilde{v} \\[3mm] 0 & 0 & 1 \end{pmatrix}$$

$$= \begin{pmatrix} \cos\alpha & -\sin\alpha & k_1 \\ \sin\alpha & \cos\alpha & k_2 \\ 0 & 0 & 1 \end{pmatrix}^{-1} \rho(u, v, u_s, \ldots)$$

which is the equivariance of a right frame for a right action.

Exercise 4.1.7 Show that taking the usual Euclidean action in the above example results in the inverse of the frame calculated above, with the same normalisation equations. Verify its equivariance.

Figure 4.5 An orthonormal vector 'frame' attached to a curve in the plane.

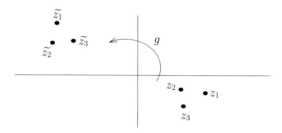

Figure 4.6 Euclidean motion on sets of points in the plane.

Example 4.1.6 provides a nice illustration of the relationship between the Fels–Olver frame and Cartan's frame of vectors at each point of the curve. If we look at the columns of the rotation part of the matrix form of ρ, we see that they form the unit tangent and unit normal of the curve, depicted as e_1 and e_2 in Figure 4.5. Cartan's frame consists of this pair, as well as the point of the curve at which they sit. The rotation angle of our frame is that required to rotate the vector pair $((1, 0), (0, 1))$ to the vector pair (e_1, e_2).

Example 4.1.8 Consider now Euclidean motion on sets of points in the plane, see Figure 4.6. Let the point z_i have coordinates (x_i, y_i). Again, let us consider the inverse action,

$$\begin{pmatrix} \widetilde{x}_i \\ \widetilde{y}_i \end{pmatrix} = \begin{pmatrix} \cos\theta & \sin\theta \\ -\sin\theta & \cos\theta \end{pmatrix} \begin{pmatrix} x_i - a \\ y_i - b \end{pmatrix}.$$

Taking the normalisation equations,

$$\widetilde{x}_1 = 0, \qquad \widetilde{y}_1 = 0, \qquad \widetilde{y}_2 = 0$$

and solving for the three group parameters, we obtain

$$a = x_1, \qquad b = y_1, \qquad \theta = \arctan\left(\frac{y_2 - y_1}{x_2 - x_1}\right). \tag{4.6}$$

We leave it to the reader to define the domain of validity of the frame.

Remark 4.1.9   The frame given by equation (4.6) works regardless of the number of points in the plane we consider. In particular, it can be used in the study of Euclidean motions on piecewise linear curves in the plane.

Remark 4.1.10   Actions on sets of points in M is equivalent to considering the product action on $M \times \cdots \times M$, see Section 1.3.2. Further examples appear in Section 4.7.

In the above examples, we have looked at two different actions induced from the standard Euclidean group action on the plane, one on curves and one on sets of points. It is not possible to define a frame for the standard action on the plane itself. One of the requirements of the construction of the frame given at the start of the chapter is that the group orbits foliate the space, and that these orbits have the same dimension as the group. Since the Euclidean group has dimension three and the plane has dimension two, this is impossible. Thus, it is necessary to find a larger space on which to act.

The examples and exercises in this and subsequent sections may give the impression that one can always solve the normalisation equations for the frame in explicit detail. Unfortunately, this is not true in general, although it is surprising just how often one *can* solve them. As we go through the theory and applications of moving frames, we will always be asking, how much information can we obtain *without* solving for the frame. The answer is: a great deal, as we shall see.

4.2 Transversality and the converse to Theorem 4.1.3

Given a free and regular group action on Ω, we constructed above moving frames in the neighbourhood of any point. Let us now consider the converse: if a moving frame exists, is the group action free and regular?

Suppose we have a smooth equivariant map $\rho : M \to G$. We first show the action is free. Let $h \in G$ be an element of the isotropy group of z. Then, for example, if we have a left frame for a left action, $h\rho(z) = \rho(h * z) = \rho(z)$. Multiplying these equations on the right by $\rho(z)^{-1}$ yields $h = e$, the identity element. Similar remarks hold for the other cases.

Next we consider the conditions for regularity. Since by hypothesis the action is free and smooth, the orbits foliate the space by the Frobenius Theorem, Theorem 3.1.24. The cross section to the orbits for the moving frame as we constructed above can be described as the set $\mathcal{K} = \{z \in M \mid \rho(z) = e\}$. This set is non-empty, since if $\rho(z) = g$ say, then $g^{-1} * z \in \mathcal{K}$. Further, each orbit intersects with $\mathcal{K}$ at most once; if not, there exists $z \in \mathcal{K}$ and $g \in G$, $g \neq$

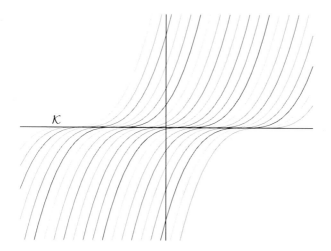

Figure 4.7 Group orbits all tangent to $\mathcal{K}$; a nowhere transverse frame.

e, such that $g * z \in \mathcal{K}$. But then by definition, $e = \rho(g * z) = g\rho(z) = g$, a contradiction.

Finally, we need to show there exists a *transverse* cross section to the orbits. To show the orbits are transverse to $\mathcal{K}$, we need to demonstrate that for any $z \in \mathcal{K}$,

$$T_z\mathcal{K} + T_z\mathcal{O}(z) = T_zM,$$

see Definition 1.4.9. Unfortunately, this is not true in general for the $\mathcal{K}$ given above.

Example 4.2.1 Consider the action of $(\mathbb{R}, +)$ on the plane whose orbits consist of the cubic curves depicted in Figure 4.7. The action is $\widetilde{x} = x + \epsilon$, $\widetilde{y} = (\epsilon + y^{1/3})^3$. Now let $\rho(x, y) = -y^{1/3}$, obtained by solving the normalisation equation $\widetilde{y} = 0$. This is a right equivariant mapping since $\rho(\widetilde{x}, \widetilde{y}) = -\widetilde{y}^{1/3} = -(\epsilon + y^{1/3}) = \rho(x, y) - \epsilon$. The set $\mathcal{K} = \{(x, y) \mid \rho(x, y) = 0\} = \{(x, 0)\}$, is the x-axis, to which every orbit is tangent, not transverse. However, it is apparent that a transverse cross section exists arbitrarily close to $\mathcal{K}$.

Arbitrary surfaces can always be deformed *locally* to be transverse to a foliation. The foliation in Figure 4.8 is given by the action $\widetilde{x} = \exp(-t)x$, $\widetilde{y} = \exp(t)y$, $t \in \mathbb{R}$, which is both free and regular on $\mathbb{R}^2 \setminus \{(0, 0)\}$. The circle $\mathcal{C}$ drawn in the figure can be deformed so that near any point on the deformed circle, the orbits are transverse to it, but this cannot be achieved simultaneously for all points. The inability to have simultaneous transversality is due to the existence of orbits arbitrarily close to $\mathcal{C}$ that do not intersect with it, no matter

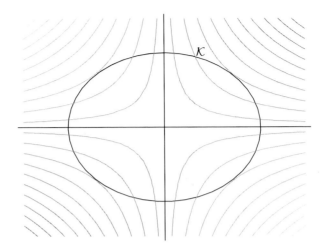

Figure 4.8 The cross section $\mathcal{K}$ cannot be deformed so as to be transverse to the foliation simultaneously at every point.

how $\mathcal{C}$ is deformed. Note that this 'cross section' does not define a global frame as most orbits intersect it either twice or not at all, but on certain open sets of $\mathbb{R}^2 \setminus \{(0, 0)\}$ it does define a local frame.

A proper discussion of the existence, at least locally, of a transverse cross section would take us too far afield into Differential Topology. However, the key words are hopefully clear: transversality is *generic*, which means that we can deform the cross section to one that is transverse, at least *locally*.

Looking more closely at the use of normalisation equations to define a frame, we see that we must solve

$$\psi_k(g * z) = 0, \qquad k = 1, \dots, r$$

for the r group parameters that describe the Lie group near its identity element. Writing the solution to these equations as $g = \rho(z)$ which gives the frame ρ, requires the application of the inverse function theorem, for which a necessary condition is that the derivative of the map $\psi : G \to \mathbb{R}^r$, where $\Psi = (\psi_1, \dots, \psi_r)$ and z is regarded as a multidimensional parameter, has non-zero determinant on the cross section. 'Non-zero determinant' is an open condition on z, and thus Ψ will typically be invertible on an open piece of the cross section $\{z \mid \Psi(z) = 0\}$: in general, we will only be able to define frames locally by this method. The right equivariance of the frame is a consequence of the uniqueness of the solution that the inverse function theorem provides. Indeed, since both $\rho(h * z)$ and $\rho(z)h^{-1}$ solve $\psi_k(g * (h * z)) = 0$, $k = 1, \dots, r$, uniqueness implies they must be equal.

However, transversality requires more than solvability of the normalisation equations, as Example 4.2.1 shows.

Why is transversality so important and desirable? Look again at Figure 4.1 at the start of this chapter. If $\mathcal{K}$ is transverse to the orbits, then ρ defines local coordinates as follows. Since the group element that sends z to $\mathcal{K}$ is unique, each $\mathcal{O}(z) \cap U$ can be given the same coordinates as that of the group about its identity element. Moreover, the element $\{k\} = \mathcal{O}(z) \cap \mathcal{K} = \rho(z) \cdot z$ tells you which orbit you are on. Thus, assuming a right frame ρ for a left action, the map

$$\varphi : U \to G \times \mathcal{K}, \qquad z \mapsto (\rho(z), \rho(z) * z) \tag{4.7}$$

is invertible on its image. Transversality guarantees the derivative of φ is also invertible, and so is a genuine coordinate transformation, or diffeomorphism. We will be using these coordinates in just about every application in this book.

If the frame is not transverse, then the map φ given in equation (4.7) will not be a diffeomorphism. Consider the frame in Example 4.2.1. Since $\mathcal{K}$ is the x-axis, it suffices to take the x-component of $\rho(z) \cdot z$, so that

$$\varphi(x, y) = (-y^{1/3}, x - y^{1/3}).$$

The Jacobian map of φ is easily seen to be non-invertible when $y = 0$, that is, on $\mathcal{K}$. Replacing the normalisation equation with $\widetilde{y} = c$ yields the frame to be $\rho(x, y) = c^{1/3} - y^{1/3}$ and $\varphi(x, y) = (c^{1/3} - y^{1/3}, x + c^{1/3} - y^{1/3})$ which has invertible Jacobian in a neighbourhood of the line $y = c$.

The Morse–Sard Theorem (see Hirsch, 1976) can be used to show that the transversality of the frame is the generic situation, and can always be achieved by altering the constants c_i in the normalisation equations $\psi_i(z) = c_i$, perhaps on a smaller neighbourhood than the one we started with. We summarise this discussion as a theorem.

Theorem 4.2.2 *Let the normalisation equations be $\psi_i(\widetilde{z}) = c_i$, and let $D_g \Psi$ denote the Jacobian of the map $\Psi = (\psi_1, \ldots, \psi_r)$ with respect to the group parameters. If the determinant*

$$D_g \Psi |_{g=e}$$

regarded as a function of z is non-zero on the locus $\mathcal{K} = \{z \mid \psi_i(z) = c_i, i = 1, \ldots, r\}$, so that the implicit function theorem may be applied, then the map $\rho : G \to M$ defined by $\psi_i(\rho(z) \cdot z) = c_i$ is a moving frame on a neighbourhood of $\mathcal{K}$. Moreover, the surface $\mathcal{K}$ can be assumed to be transverse to the group orbits for generic c_i.

In all that follows, we will assume that the orbits are transverse to the cross section that defines the frame.

Remark 4.2.3 For algebraic actions, one can dispense with the inverse function theorem, see Hubert and Kogan (2007a, 2007b) replacing its role in the theory with results from commutative algebra. This means that well-defined algorithms can be developed and algebraically certified for studying moving frames and their applications in a symbolic computing environment.

4.3 Frames for *SL(2)* actions

We now turn our attention to the three non-linear actions of $SL(2)$ in the plane, given in Chapter 1. None of these actions is free as they stand, so we need to extend the space on which they act in some way. We will use these frames in examples all through the book.

Example 4.3.1 To calculate the frame for the action (1.17), as induced on curves $(x(s), y(s))$ in the plane, we take normalisation equations,

$$\widetilde{x} = 0, \qquad \widetilde{x_s} = 1, \qquad \widetilde{x_{ss}} = 0.$$

Near the identity element, the group parameter $d = (1 + bc)/a$. Solving the normalisation equations for the three independent group parameters yields

$$a = \frac{1}{\sqrt{x_s}}, \qquad b = -\frac{x}{\sqrt{x_s}}, \qquad c = \frac{x_{ss}}{2(x_s)^{3/2}}.$$

In matrix form, the moving frame is

$$\rho(x, x_s, x_{ss}, \dots) = \begin{pmatrix} \dfrac{1}{\sqrt{x_s}} & -\dfrac{x}{\sqrt{x_s}} \\ \dfrac{x_{ss}}{2(x_s)^{3/2}} & \dfrac{2x_s^2 - xx_{ss}}{2(x_s)^{3/2}} \end{pmatrix}. \tag{4.8}$$

The choice of the positive square root is to ensure that ρ is the identity element on the cross section $\mathcal{K}$.

Since the normalisation equations, which define the cross section, include $\widetilde{x_s} = 1$, the frame is defined in a neighbourhood of $(x, y, x_s, y_s, \dots)$-space where $x_s \neq 0$. It can be seen by examining the group action on x_s,

$$\widetilde{x_s} = \frac{x_s}{(cx + d)^2},$$

that the group action leaves the coordinate slice, $x_s = 0$, invariant, and in fact is singular there (see Olver (2000) for a discussion of the singularities of group actions).

Exercise 4.3.2 Which equivariance equation is satisfied by the frame (4.8)?

Exercise 4.3.3 Suppose instead one induces the action (1.17) on curves parametrised as $(x, y(x))$ in the plane. Calculate the frame for the normalisation equations,

$$\widetilde{x} = 0, \qquad \widetilde{y_x} = 1, \qquad \widetilde{y_{xx}} = 0.$$

Explain why it is not possible to have a normalisation equation of the form $y = c$.

Exercise 4.3.4 Calculate frames for curves given as $(x(s), y(s))$ and as $(x, y(x))$ for the $SL(2)$ actions, (1.18) and (1.19). Hint: in order to have an equation for the group parameter b, it is necessary to have a normalisation equation of the form $\widetilde{x} = c$. One always chooses the simplest possible form of the constants involved, usually either 0 or 1; in this case, $\widetilde{x} = 0$ is recommended.

4.4 Invariants

One can think of the transverse moving frame, defined in U with cross section $\mathcal{K} \subset U$, as providing local coordinates, sometimes called a local trivialisation, of the manifold, as discussed in Section 4.2. For a *right* frame, the coordinate transformation is given by

$$\varphi : U \to G \times \mathcal{K}, \qquad z \mapsto (\rho(z), \rho(z) \cdot z).$$

The leaves of the foliation given by the group orbits are parametrised by the group parameters, while $\rho(z) \cdot z$ yields the element $\mathcal{O}(z) \cap \mathcal{K}$. We show in this section that $\mathcal{K}$ has coordinates given by *invariants* of the group action.

Theorem 4.4.1 *If ρ is a right frame, then the quantity $I(z) = \rho(z) \cdot z$ is an invariant of the group action.*

Proof For a left action we have

$$I(g * z) = \rho(g * z) * (g * z) = (\rho(g * z)g) * z = (\rho(z)g^{-1}g) * z$$
$$= \rho(z) * z = I(z).$$

For a right action, $\rho(z) \bullet z$ is similarly invariant; since $g \bullet (h \bullet z) = (hg) \bullet z$, we have

$$\rho(g \bullet z) \bullet (g \bullet z) = (g\rho(g \bullet z)) \bullet z = (gg^{-1}\rho(z)) \bullet z = \rho(z) \bullet z.$$

So, we write both cases simply as $\rho(z) \cdot z$. □

Definition 4.4.2 The map $z \mapsto I(z) = \rho(z) \cdot z$ is called the *invariantisation map*. Other notations in use are $\iota(z)$ and $\overline{\iota}z$.

Solving normalisation equations always yields a right frame, so that evaluating $\widetilde{z} = g \cdot z$ with $g = \rho(z)$ in parameter form yields an invariant. This is true whether we consider z in coordinates or not; even though the parity of the action is opposite for z and its coordinates, the equivariance changes to match. Thus, if $z = (z_1, z_2, \ldots, z_n)$ and the normalisation equations are $\widetilde{z}_i = c_i$ for $i = 1, \ldots, r$, where r is the dimension of the group, then the components of

$$\rho(z) \cdot z = (c_1, \ldots, c_r, I(z_{r+1}), \ldots, I(z_n))$$

where

$$I(z_k) = g \cdot z_k \big|_{g = \rho(z)}$$

are all invariants.

We note that the $I(z_k)$ are all functionally independent. For if there were an expression of the form $F(I(z_k)) \equiv 0$, then by back-substituting for the $I(z_k)$ their expressions in terms of the z_i, we would have that the z_i were functionally dependent, a contradiction.

Seeing is believing, so we turn our attention to considering the invariants for the examples considered in the previous section.

Example 4.4.3 For the rotation group in the plane, Example 4.1.4, the calculation looks like

$$g \cdot z \big|_{g = \rho(z)}$$

$$= ((\cos\theta)x - (\sin\theta)y, (\sin\theta)x + (\cos\theta)y)\big|_{\theta = \arctan(x/y)}$$

$$= (0, \sqrt{x^2 + y^2}).$$

The first component is zero because the normalisation equation is $\widetilde{x} = 0$; the frame is defined to be the group element that sends the first component of $\widetilde{z}$ to zero. The second component of $\rho(z) \cdot z$ is evidently invariant. Moreover, looking at Figure 4.3, it is clear that the intersection of an orbit passing through the point (x, y), with the positive y-axis, is the point $(0, \sqrt{x^2 + y^2})$. In this simple case, the coordinates provided by the frame are essentially the usual polar coordinates.

Example 4.4.4 Consider next the Euclidean action on sets of points in the plane, Example 4.1.8. Calculating $\rho(z) \cdot z$ means back-substituting in $g \cdot z$ the specific values of the parameters that determine the frame, equation (4.6), in the $\widetilde{z}_i = g \cdot z_i$. Doing this yields $I(z_1) = (0, 0)$, as expected, since these are the first two normalisation equations, and then

$$I(z_2) = (\sqrt{(x_2 - x_1)^2 + (y_2 - y_1)^2}, 0)$$

and

$$I(z_3) = \left(\frac{(x_3 - x_1)(x_2 - x_1) + (y_3 - y_1)(y_2 - y_1)}{\sqrt{(x_2 - x_1)^2 + (y_2 - y_1)^2}}, \right.$$

$$\left. \frac{(y_3 - y_1)(x_2 - x_1) - (x_3 - x_1)(y_2 - y_1)}{\sqrt{(x_2 - x_1)^2 + (y_2 - y_1)^2}} \right).$$

Each component of $I(z_2)$ and $I(z_3)$ is an invariant. The second component of $I(z_2)$ is zero since that is the third normalisation equation. The geometric interpretation of the non-trivial invariants are as functions of lengths, angles between two line segments that meet at a point, and areas of triangles, all of which are evidently invariant under the Euclidean group action on the plane.

Example 4.4.5 Looking at the Euclidean action on curves in the plane, Figure 4.4, Example 4.1.6, the frame in parameter space is given in equation (4.4). The coordinates are $(s, u, v, u_x, v_x, u_{xx}, \dots)$. The invariants are the components of

$$(g \cdot s, g \cdot u, g \cdot v, g \cdot u_s, g \cdot v_s, g \cdot u_{ss}, g \cdot v_{ss}, \dots) \Big|_{g = \rho}$$

$$= \left(s, 0, 0, \sqrt{u_s^2 + v_s^2}, 0, \frac{u_{ss} u_s + v_{ss} v_s}{\sqrt{u_s^2 + v_s^2}}, \frac{u_{ss} v_s - v_{ss} u_s}{\sqrt{u_s^2 + v_s^2}}, \dots \right).$$

Exercise 4.4.6 The radius of any circle in the plane is invariant under translations and rotations, and hence the radius of an osculating circle to a curve, pictured in Figure 4.9, is a Euclidean invariant. The Euclidean curvature of a curve at a point is the reciprocal of the radius of the osculating circle there. If the curve is parametrised as $(x, y(x))$, then show the curvature is

$$\kappa = \frac{y_{xx}}{(1 + y_x^2)^{3/2}}.$$

Hint: let the equation of the circle be $c(x) = \sqrt{r^2 - (x - x_c)^2} + y_c$ where r is the radius and (x_c, y_c) its centre. Then there are three equations for the three unknowns, r, x_c and y_c, namely $y_0 = y(x_0) = c(x_0)$, $y_x = c_x$ and $y_{xx} = c_{xx}$ evaluated at the point of contact (x_0, y_0).

Exercise 4.4.7 Consider the inverse Euclidean action on curves in the plane, but this time described by $(x, y(x))$, so that the action takes place in $(x, y, y_x, y_{xx}, \dots)$-space. Let the normalisation equations be

$$\tilde{x} = 0, \qquad \tilde{y} = 0, \qquad \tilde{y}_x = 0.$$

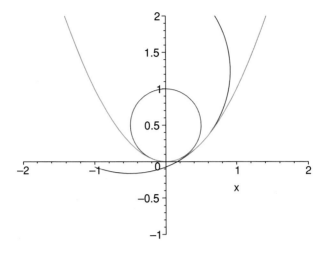

Figure 4.9 Two osculating circles, at $x = 0$ (blue) and at $x = 0.5$ (in black), on the parabola (x, x^2) drawn in green. Note the osculating circles do not necessarily lie on only one side of the curve.

Show that the frame is given by

$$a = x, \qquad b = y, \qquad \theta = \arctan y_x$$

and that

$$\widetilde{y_{xx}}\Big|_{g=\rho} = I(y_{xx}) = \frac{y_{xx}}{(1 + y_x^2)^{3/2}}$$

which is the Euclidean curvature defined in the previous exercise. Find $I(y_{xxx})$ and $I(y_{xxxx})$ explicitly.

Definition 4.4.8 For any prolonged action in $(x_i, u^\alpha, u_K^\alpha)$-space, the specific components of $I(z)$, the invariantised jet coordinates, are denoted

$$J_i = I(x_i) = \widetilde{x}_i\Big|_{g=\rho(z)}, \qquad I_K^\alpha = I(u_K^\alpha) = \widetilde{u}_K^\alpha\Big|_{g=\rho(z)} \tag{4.9}$$

which is the original Fels and Olver notation. More recently, some authors have denoted these as $\iota(x_i)$, $\iota(u_K^\alpha)$ or $\bar{\iota}x_i$, $\bar{\iota}u_K^\alpha$.

Explicit expressions for these invariants in terms of the original variables can often be obtained even when the frame is not known explicitly. This is because the frame dependent invariants $I(z_k)$ may be related to historically known invariants using the Fels–Olver–Thomas Replacement Theorem (Fels and Olver, 1999, Theorem 10.3) which states the following.

Theorem 4.4.9 *If $f(z)$ is an invariant then*

$$f(z) = f(I(z)).$$

Proof Substitute $g = \rho(z)$ into $f(z) = f(g \cdot z)$. ☐

An important corollary is the following.

Theorem 4.4.10 (Replacement Theorem) *Any invariant is a function of the $I(z_k)$. In particular, the set $\{J_i, I^\alpha, I_K^\alpha\}$, defined in Definition 4.4.8, is a complete set of differential invariants for a prolonged action.*

Example 4.4.11 To show the power of the Replacement Theorem, consider the action induced on curves by the group of Euclidean motions in 3-space. We take a curve parametrised as $\mathbf{u} = (u(s), v(s), w(s))$. The action leaves the curve parameter s invariant, so $\widetilde{s} = s$, and sends $\widetilde{\mathbf{u}} = R\mathbf{u} + \mathbf{a}$ where $R \in SO(3)$ and $\mathbf{a} \in \mathbb{R}^3$ is the constant vector of translation. Let the normalisation equations be

$$\widetilde{u} = \widetilde{v} = \widetilde{w} = 0, \qquad \widetilde{v_s} = \widetilde{w_s} = 0, \qquad \widetilde{w_{ss}} = 0.$$

Using the notation of Definition 4.4.8, we have $I^u = I^v = I^w = 0, I_1^v = I_1^w = 0$ and $I_{11}^w = 0$. We do not solve for the frame explicitly in order to calculate the other I_K^α. Instead, we apply the Replacement Theorem to obtain the I_K^α in terms of historically known invariants. Thus, from the invariant $|\mathbf{u}_s|$ where $\mathbf{u}_s = (u_s, v_s, w_s)$, we obtain

$$\sqrt{u_s^2 + v_s^2 + w_s^2} = \sqrt{(I_1^u)^2 + (I_1^v)^2 + (I_1^w)^2} = |I_1^u|,$$

applying the normalisation equations. This solves for I_1^u up to a sign. Another known invariant is the curvature, $\kappa = |\mathbf{u}_s \times \mathbf{u}_{ss}|/|\mathbf{u}_s|^{3/2}$. Applying the Replacement Theorem and the normalisation equations yields an expression for I_{11}^v in terms of κ and $|\mathbf{u}_s|$. Finally, applying the method to the invariant determinant $\tau = (\mathbf{u}_s \times \mathbf{u}_{ss}) \cdot \mathbf{u}_{sss}/\kappa^2$ which defines τ, the torsion of the curve, yields an expression for I_{111}^w. The curvature and torsion form a complete set of invariants for the Euclidean action on curves in 3-space; any other invariant is a function of these and their derivatives with respect to s. Thus, the remaining I_K^α can all be obtained by differentiation and replacement. Further details appear in Section 5.4.

The Replacement Theorem implies that expressing any invariant in terms of the I_K^α is achieved by simple substitution. This process is called *invariantisation*. In a computer algebra environment, invariantisation is achieved by substitution of the normalised invariants followed by simplification with respect to the normalisation equations. For a discussion of the subtle issues that arise in this context we refer to Mansfield (2001).

Our final example in this section will prove important in the applications in later chapters.

Example 4.4.12 If we consider left multiplication on any matrix Lie group, prolonged to the tangent space TG, we obtain

$$G \times TG \to TG, \qquad h * (g, g') \mapsto (hg, hg').$$

Taking the normalisation equation $\widetilde{g} = e$, then the frame is given explicitly by

$$\rho(g, g') = g^{-1}.$$

Moreover the components of

$$I(g, g') = g^{-1} g' \in \mathfrak{g}$$

are invariants under left multiplication.

4.5 Invariant differentiation

This section relates purely to moving frames for prolonged actions. We investigate not only *invariant differential operators* but the formulae for the invariant differentiation of the invariants I_K^{α}.

The method for obtaining invariant differential operators is similar to that of obtaining differential invariants.

Example 4.5.1 A simple example is provided by the scaling action in the plane,

$$\widetilde{x} = \lambda x, \qquad \widetilde{u} = u/\lambda, \qquad \lambda > 0$$

and prolonged to curves parametrised as $(x, u(x))$. With the normalisation equation $\widetilde{x} = 1$ on the domain $\{(x, u, u_x, \dots) \mid x > 0\}$, the frame is $\lambda = 1/x$.

The invariant differential operator is then

$$\mathcal{D} = \frac{\mathrm{d}}{\mathrm{d}\widetilde{x}}\bigg|_{\lambda=1/x} = \frac{\mathrm{d}x}{\mathrm{d}\widetilde{x}} \frac{\mathrm{d}}{\mathrm{d}x}\bigg|_{\lambda=1/x} = x\frac{\mathrm{d}}{\mathrm{d}x}.$$

To see this is invariant, note that

$$\widetilde{x}\frac{\mathrm{d}}{\mathrm{d}\widetilde{x}} = \widetilde{x}\frac{\mathrm{d}x}{\mathrm{d}\widetilde{x}} \frac{\mathrm{d}}{\mathrm{d}x} = x\frac{\mathrm{d}}{\mathrm{d}x}.$$

Evaluating $\widetilde{u}$ on the frame leads to the invariant $xu = \widetilde{u}|_{\lambda=1/x}$. Since

$$\frac{\widetilde{\mathrm{d}^k u}}{\mathrm{d}x^k} = \lambda^{-(k+1)}\frac{\mathrm{d}^k u}{\mathrm{d}x^k},$$

we have invariants

$$I_k = \left.\widetilde{\frac{\mathrm{d}^k u}{\mathrm{d}x^k}}\right|_{\lambda=1/x} = x^{k+1}\frac{\mathrm{d}^k u}{\mathrm{d}x^k}$$

where we have suppressed the index u on $I_k^u = I_k$ as there is only one dependent variable. It is interesting to compare I_1 with $\mathcal{D}I_0$, and more generally, I_{k+1} with $\mathcal{D}I_k$. It is trivial to verify

$$\mathcal{D}I_k = (k+1)I_k + I_{k+1},$$

and thus we see that even though

$$\frac{\mathrm{d}}{\mathrm{d}x}\frac{\mathrm{d}^k u}{\mathrm{d}x^k} = \frac{\mathrm{d}^{k+1}u}{\mathrm{d}x^{k+1}},$$

the same is not true for their invariantised counterparts,

$$\mathcal{D}I_k \neq I_{k+1}. \tag{4.10}$$

Since to obtain $\mathcal{D}I_k$ we first invariantise $\mathrm{d}^k u/\mathrm{d}x^k$ and then differentiate, while to obtain I_{k+1} we first differentiate and then invariantise, equation (4.10) says, 'differentiation and invariantisation do not commute'.

Definition 4.5.2 A set of distinguished *invariant operators* is defined by evaluating the transformed total differential operators on the frame. They are

$$\mathcal{D}_j = \left.\widetilde{D}_j\right|_{g=\rho(z)},$$

where the $\widetilde{D}_j$ are given in Chapter 1, equation (1.35).

By an argument similar to the Replacement Theorem, Theorem 4.4.9, any invariant differential operator can be written in terms of the $\mathcal{D}_j$ and the symbolic invariants $J_i = \iota(x_i)$, $I^\alpha = \iota(u^\alpha)$ and $I_K^\alpha = \iota(u_K^\alpha)$. The $\mathcal{D}_j$ are linear derivations on functions of J_i, I^α and I_K^α, that is, they are linear operators and the product rule holds. As we shall see, in general,

$$[\mathcal{D}_j, \mathcal{D}_k] \neq 0$$

and as Example 4.5.1 shows,

$$\mathcal{D}_j I_K^\alpha = \left.\widetilde{D}_j\right|_{g=\rho(z)}\left.\widetilde{u_K^\alpha}\right|_{g=\rho(z)} \neq \left.\widetilde{D}_j\widetilde{u_K^\alpha}\right|_{g=\rho(z)} = \left.\widetilde{u_{Kj}^\alpha}\right|_{g=\rho(z)} = I_{Kj}^\alpha.$$

This motivates the following definition.

Definition 4.5.3 The *correction terms* N_{ij} and M_{Kj}^α are defined by

$$\mathcal{D}_j J_i = \delta_{ij} + N_{ij}, \qquad \mathcal{D}_j I_K^\alpha = I_{Kj}^\alpha + M_{Kj}^\alpha, \tag{4.11}$$

where δ_{ij} is the Kronecker delta.

It follows from their definition that the invariants I_K^α are left unchanged by permutations within their index K. The correction terms, however, are *not* invariant under permutations in their index, because the operators do not commute.

In order to understand the formulae for the correction terms, we need to recall the infinitesimals of the group action in Section 1.6. If the parameters of the group in a neighbourhood of the identity element are $a_1, \ldots, a_r$ where r is the dimension of the group, then the group infinitesimals with respect to these parameters are

$$\xi_i^j = \frac{\partial \widetilde{x}_j}{\partial a_i}\Big|_{g=e}, \qquad \phi_{K,i}^\alpha = \frac{\partial \widetilde{u_K^\alpha}}{\partial a_i}\Big|_{g=e}.$$

These are functions of $x_i, i = 1, \ldots, p, u^\alpha, \alpha = 1, \ldots, q$, and u_K^α.

Definition 4.5.4 Given $\xi_i^j = \xi_i^j(x, u^\beta)$, define $\xi_i^j(I) = \xi_i^j(J, I^\beta)$ and similarly $\phi_K^\alpha(I) = \phi_K^\alpha(J, I^\beta, I_L^\beta)$, that is, where the arguments have been invariantised.

Theorem 4.5.5 *There exists a $p \times r$ correction matrix* $\mathbf{K}$ *such that*

$$N_{kj} = \sum_{\ell=1}^r \mathbf{K}_{j\ell} \xi_\ell^k(I), \qquad M_{Kj}^\alpha = \sum_{\ell=1}^r \mathbf{K}_{j\ell} \phi_{K,\ell}^\alpha(I) \qquad (4.12)$$

where ℓ is the index for the group parameters and $r = \dim(G)$. If the action on the base space is left and the frame is right, then $\mathbf{K}$ *is given by*

$$\mathbf{K}_{j\ell} = \widetilde{D}_j \rho_\ell(\widetilde{z})\Big|_{g=\rho(z)} = ((T_e R_\rho)^{-1} \mathcal{D}_j \rho)_\ell \qquad (4.13)$$

where $\rho = (\rho_1, \ldots, \rho_r)^T$ is in parameter form and $R_\rho : G \to G$ is right multiplication by ρ.

Proof To show the result for the M_{Kj}^α, we apply the chain rule to $\widetilde{D}_j I(z)$ evaluated at $g = \rho(z)$. The proof for the N_{kj} is entirely similar. Writing

$$\widetilde{u_K^\alpha} = F_K^\alpha(a_1, \ldots, a_r, x, u^\beta, \ldots)$$

where the a_i are the group parameters (near e), we have by definition that

$$\widetilde{D}_j F_K^\alpha = F_{Kj}^\alpha.$$

Assuming a right frame, we have

$$I_k^\alpha = F_K^\alpha(\rho_1, \ldots, \rho_r, x, u^\beta, \ldots).$$

If the frame is left, then one must use ρ^{-1} in place of ρ in F_K^α to obtain the invariant I_K^α; we leave this case to the reader.

Applying the chain rule, we have

$$\widetilde{D}_j\Big|_{g=\rho} I_K^\alpha = \sum_{\ell=1}^r \left(\frac{\partial F_K^\alpha}{\partial g_\ell}\right)\Big|_{g=\rho} \left(\widetilde{D}_j \rho_\ell\right)\Big|_{g=\rho} + \left(\widetilde{D}_j F_K^\alpha\right)\Big|_{g=\rho}. \quad (4.14)$$

The second summand of equation (4.14) is I_{Kj}^α by definition. By Theorem 3.2.37, we have

$$\left(\frac{\partial F_K^\alpha}{\partial g_1}(g,z), \ldots, \frac{\partial F_K^\alpha}{\partial g_r}(g,z)\right)^T\Big|_g = (T_e R_g)^{-T} \Phi_K^\alpha(g \cdot z),$$

where $z = (x, u^\beta, \ldots)$ and by definition

$$\left(\frac{\partial F_K^\alpha}{\partial g_1}(g,z), \ldots, \frac{\partial F_K^\alpha}{\partial g_r}(g,z)\right)^T\Big|_{g=e} = \Phi_K^\alpha(z),$$

the vector of infinitesimals of the group action on the u_K^α coordinate with respect to the group parameters $(a_1, \ldots, a_r)$.

Hence the first summand of equation (4.14) is

$$\sum_{\ell=1}^r \Phi_{K,\ell}^\alpha(I) \left(T_e R_\rho^{-1} \mathcal{D}_j \rho\right)_\ell$$

where $\Phi_{K,\ell}^\alpha(I)$ is defined in Definition 4.5.4 above. This proves one of the expressions for **K**. Finally, we note that $\rho(g * z) = \rho(z)g^{-1}$ implies

$$\widetilde{D}_j \rho(g * z) = \widetilde{D}_j R_{g^{-1}}(\rho(z)) = T_\rho R_{g^{-1}} \widetilde{D}_j \rho(z)$$

by the chain rule. Evaluating this at $g = \rho$ and using the result of Exercise 3.2.38, we have proved the equality of the two expressions for the components of **K**. $\qquad \square$

The rows of **K** will take on additional significance in Chapter 5.

It is of great computational importance in the applications that the matrix **K** can be calculated without explicit knowledge of the frame. All that is required are the normalisation equations $\{\psi_\lambda(z) = 0, \lambda = 1, \ldots, r\}$ and the infinitesimals. Suppose the n variables actually occurring in the $\psi_\lambda(z)$ are $\zeta_1, \ldots, \zeta_n$; typically m of these will be independent variables and $n - m$ of them will be dependent variables and their derivatives. Define **T** to be the invariant $p \times n$ total derivative matrix

$$\mathbf{T}_{ij} = I\left(\frac{D}{Dx_i}\zeta_j\right).$$

Also, let Φ denote the $r \times n$ matrix of infinitesimals with invariantised arguments,

$$\Phi_{ij} = \left(\frac{\partial(g \cdot \zeta_j)}{\partial g_i}\Big|_{g=e}\right)(I).$$

Furthermore, define **J** to be the $n \times r$ transpose of the Jacobian matrix of the left hand sides of the normalisation equations $\psi_1, \ldots, \psi_r$, with invariantised arguments, that is

$$\mathbf{J}_{ij} = \frac{\partial \psi_j(I)}{\partial I(\zeta_i)}.$$

Using the above defined matrices, which are easily calculated, the correction matrix can be obtained as follows.

Theorem 4.5.6 *The correction matrix **K**, which provides the error terms in the process of invariant differentiation in Theorem 4.5.5 is given by*

$$\mathbf{K} = -\mathbf{T}\mathbf{J}(\Phi\mathbf{J})^{-1},$$

*where **T**, **J** and Φ are defined above.*

Proof We compute the invariantisation of the equations

$$D_i \psi_\lambda(\rho(z) \cdot \zeta) = 0 \qquad (4.15)$$

where $\zeta = \zeta_1, \ldots, \zeta_n$ are the actual arguments of the ψ_ℓ. The invariantised normalisation equations are functions of both the variables ζ_l and the coordinates of the frame $\rho_j(z)$. Since the latter depend on the former we have to be careful. We separate the different dependences by writing $\psi_p(\rho(z) \cdot \zeta) = \Psi_p(\zeta, \rho(z))$. Here the ψ are functions of n variables, whereas the Ψ depend on $n + r$ variables. Thus from equation (4.15) we obtain

$$\sum_{j=1}^{r} D_i \rho_j(z) \frac{\partial \Psi_\lambda(\zeta, \rho(z))}{\partial \rho_j(z)} + \sum_{l=1}^{n} D_i \zeta_l \frac{\partial \Psi_\lambda(\zeta, \rho(z))}{\partial \zeta_l} = 0.$$

We use the chain rule once more and write

$$\frac{\partial \Psi_\lambda(\zeta, \rho(z))}{\partial \rho_j(z)} = \sum_{l=1}^{n} \frac{\partial \rho(z) \cdot \zeta_l}{\partial \rho_j(z)} \frac{\partial \psi_\lambda(\rho(z) \cdot \zeta)}{\partial \rho(z) \cdot \zeta_l}.$$

The theorem is proved by invariantisation of the different terms, that is, replace z by $\widetilde{z}$ (ζ by $\widetilde{\zeta}$) and evaluate at $g = \rho(z)$. □

To calculate **K** in practice, it is easier to use labelled rows and columns for the intermediate matrices Φ, **J** and **T** rather than indices.

Example 4.5.7 Consider the prolongation of the $SL(2)$ action on $(x, t, u(x, t))$ space,

$$\widetilde{x} = x, \qquad \widetilde{t} = t, \qquad \widetilde{u} = \frac{au + b}{cu + d}.$$

The table of infinitesimals is

$$
\begin{array}{c}
 & \begin{array}{cccccc} x & t & u & u_x & u_t & u_{xx} \quad\cdots \end{array} \\
\begin{array}{c} a \\ b \\ c \end{array}
\begin{pmatrix}
0 & 0 & 2u & 2u_x & 2u_t & 2u_{xx} & \cdots \\
0 & 0 & 1 & 0 & 0 & 0 & \cdots \\
0 & 0 & -u^2 & -2uu_x & -2uu_t & -2u_x^2 - 2uu_{xx} & \cdots
\end{pmatrix}.
\end{array}
$$

If we take the normalisation equations $\widetilde{u} = 0$, $\widetilde{u}_x - 1 = 0$ and $\widetilde{u}_{xx} = 0$, then the ψ_ℓ are $\psi_1(u, u_x, u_{xx}) = u$, $\psi_2(u, u_x, u_{xx}) = u_x - 1$ and $\psi_3(u, u_x, u_{xx}) = u_{xx}$, so the arguments ζ_i are u, u_x and u_{xx}, and the invariantised normalisation equations are $I^u = 0$, $I_1^u = 1$ and $I_{11}^u = 0$. Thus, selecting the appropriate columns of the table of infinitesimals and invariantising, we obtain

$$
\Phi =
\begin{array}{c}
 & \begin{array}{ccc} u & u_x & u_{xx} \end{array} \\
\begin{array}{c} a \\ b \\ c \end{array}
\begin{pmatrix}
0 & 2 & 0 \\
1 & 0 & 0 \\
0 & 0 & -2
\end{pmatrix}.
\end{array}
$$

The Jacobian matrix is the identity matrix,

$$
\mathbf{J} =
\begin{array}{c}
 & \begin{array}{ccc} \psi_1(I) & \psi_2(I) & \psi_3(I) \end{array} \\
\begin{array}{c} I^u \\ I_1^u \\ I_{11}^u \end{array}
\begin{pmatrix}
1 & 0 & 0 \\
0 & 1 & 0 \\
0 & 0 & 1
\end{pmatrix},
\end{array}
$$

and

$$
\mathbf{T} =
\begin{array}{c}
 & \begin{array}{ccc} u & u_x & u_{xx} \end{array} \\
\begin{array}{c} x \\ t \end{array}
\begin{pmatrix}
1 & 0 & I_{111}^u \\
I_2^u & I_{12}^u & I_{112}^u
\end{pmatrix}.
\end{array}
$$

Hence

$$
-\mathbf{K} =
\begin{array}{c}
 & \begin{array}{ccc} a & b & c \end{array} \\
\begin{array}{c} x \\ t \end{array}
\begin{pmatrix}
0 & 1 & -\frac{1}{2} I_{111}^u \\
\frac{1}{2} I_{12}^u & I_2^u & -\frac{1}{2} I_{112}^u
\end{pmatrix}.
\end{array}
$$

Using $\mathbf{K}$ to calculate symbolic invariant derivatives, we have using the formula (4.12), that

$$
\mathcal{D}_x I_2^u = I_{12}^u +
\begin{array}{ccc}
\begin{array}{ccc} a & b & c \end{array} \\
\left(0 \quad -1 \quad \frac{1}{2} I_{111}^u \right)
\end{array}
\begin{array}{c}
\begin{pmatrix} 2 I_2^u \\ 0 \\ 0 \end{pmatrix}
\begin{array}{c} a \\ b \\ c \end{array}
\end{array}
= I_{12}^u,
$$

where we use the row of **K** corresponding to x since we are calculating $\mathcal{D}_x$ of something, and the column from the table of infinitesimals corresponding to u_t since we are calculating a derivative of $I_2^u = I(u_t)$.

Exercise 4.5.8 Example 4.5.7 continued. By calculating the infinitesimals for $u_{xt}, u_{xxx}, u_{xxxx}$ and u_{xxt}, show that

$$
\begin{aligned}
\mathcal{D}_x I_{12}^u &= I_{112}^u \\
\mathcal{D}_x I_{112}^u &= I_{1112}^u - 2I_{12}^u I_{111}^u \\
\mathcal{D}_x I_{111}^u &= I_{1111}^u \\
\mathcal{D}_t I_{111}^u &= I_{1112}^u - I_{12}^u I_{111}^u.
\end{aligned}
$$

Exercise 4.5.9 Consider the third of the $SL(2)$ actions in Example 1.2.14. The table of infinitesimals is given in Exercise 1.6.15. Using the normalisation equations $\widetilde{x} = 0$, $\widetilde{y} = \lambda$, $\widetilde{y}_x = 0$, where λ is an arbitrary constant, show the **K** matrix is

$$
-\mathbf{K} = \begin{array}{c} \\ x \\ t \end{array}
\begin{array}{c} a \\ \left(\begin{array}{c} \dfrac{3}{2\lambda^2} I_{11}^y \\[2mm] -\dfrac{1}{2\lambda} I_2^y + \dfrac{3}{2\lambda^2} I_{12}^y \end{array} \right. \end{array}
\begin{array}{c} b \\ \begin{array}{c} 1 \\[2mm] 0 \end{array} \end{array}
\begin{array}{c} c \\ \left. \begin{array}{c} \dfrac{1}{2\lambda} I_{11}^y \\[2mm] \dfrac{1}{2\lambda} I_{12}^y \end{array} \right). \end{array}
$$

Exercise 4.5.10 Using the code written for Exercise 1.6.21, implement the formula for both **K** given in Theorem 4.5.6, and hence the symbolic differentiation formulae, in Theorem 4.5.5. Your input will now include the normalisation equations. You will need as a subroutine an invariantisation procedure, to calculate $\phi_K^\alpha(I)$ from ϕ_K^α. You also need to be able to simplify expressions with respect to the normalisation equations with invariantised arguments, that is, the $\psi_i(I)$.

An examination of the formulae for the M_{Kj}^α shows that even when $|K|$ is less than the order of the normalisation equations, M_{Kj}^α may contain terms whose order is up to one more than the order of the normalisation equations. Further, the M_{Kj}^α will cancel the I_{Kj}^α term if I_K^α is a normalised invariant, that is, if $\widetilde{u_K^\alpha} = c$ is a normalisation equation. The consequences of the existence and formulae for the M_{Kj}^α are discussed in Chapter 5. In particular, examples there will show that the invariants I_K^α are *not* necessarily the result of differentiating some other quantity.

Since the invariant operators do not in general commute, $M_{jK}^\alpha \neq M_{Kj}^\alpha$. For this reason, the index on the M_{jK}^α is *not* of the form of an element in $\mathbb{N}^p$; it is important to keep track of the order in which the differential operators are applied.

The commutators of the invariant derivative operators can be calculated using only the **K** matrix and the infinitesimals of the group action. The following formula is taken from Fels and Olver (1999), Equation 13.12. Denote the invariantised derivatives of the infinitesimals ξ by

$$\Xi^k_{li} = \widetilde{D}_i \xi^k_l(\widetilde{z})\Big|_{g=\rho(z)}.$$

Then we have

$$[\mathcal{D}_i, \mathcal{D}_j] = \sum_k A^k_{ij} \mathcal{D}_k, \qquad A^k_{ij} = \sum_{l=1}^r \mathsf{K}_{jl} \Xi^k_{li} - \mathsf{K}_{il} \Xi^k_{lj}. \tag{4.16}$$

The proof of equation (4.16) uses the formula for the Lie bracket of vector fields and $\partial \widetilde{x}_i / \partial g_j = \xi^i_j(\widetilde{x}_i)$. It is not necessary to implement it if you have implemented the symbolic invariant differentiation formulae. Simply calculate $[\mathcal{D}_i, \mathcal{D}_j] I^w$ where w is an invariant dependent variable introduced for the purpose. The coefficient of I^w_k will be A^k_{ij}.

Exercise 4.5.11 Consider the action of $SE(2)$ on the plane,

$$\begin{pmatrix} \widetilde{x} \\ \widetilde{y} \end{pmatrix} = \begin{pmatrix} \cos\theta & \sin\theta \\ -\sin\theta & \cos\theta \end{pmatrix} \begin{pmatrix} x - a \\ y - b \end{pmatrix}$$

and let $u = u(x, y)$ be an invariant. Take the normalisation equations $\widetilde{x} = 0$, $\widetilde{y} = 0$, $\widetilde{u}_x = 0$. Show that

$$\mathcal{D}_x = \frac{u_y}{\sqrt{u_x^2 + u_y^2}} \frac{D}{Dx} - \frac{u_x}{\sqrt{u_x^2 + u_y^2}} \frac{D}{Dy}, \quad \mathcal{D}_y = \frac{u_x}{\sqrt{u_x^2 + u_y^2}} \frac{D}{Dx} + \frac{u_y}{\sqrt{u_x^2 + u_y^2}} \frac{D}{Dy}$$

and that

$$[\mathcal{D}_x, \mathcal{D}_y] = \frac{I^u_{11}}{I^u_2} \mathcal{D}_x + \frac{I^u_{12}}{I^u_2} \mathcal{D}_y.$$

Calculate I^u_{11}, I^u_{12} and I^u_2 explicitly and verify the result both directly and using the symbolic differentiation formulae. Note the error matrix for these normalisation equations is

$$\mathsf{K} = \begin{matrix} & a & b & \theta \\ x & \\ y & \end{matrix} \begin{pmatrix} -1 & 0 & I^u_{11}/I^u_2 \\ 0 & -1 & I^u_{12}/I^u_2 \end{pmatrix}.$$

4.5.1 Invariant differentiation for linear actions of matrix Lie groups

If the action of a matrix Lie group $G \subset GL(n, \mathbb{R})$ is the standard left multiplication on $\mathbb{R}^n$, or more generally is the standard affine action of $G \ltimes \mathbb{R}^n$ on $\mathbb{R}^n$, it is not necessary to put coordinates on G in order to calculate the error terms M^α_{Kj} in $\mathcal{D}_j I^\alpha_K = I^\alpha_{Kj} + M^\alpha_{Kj}$.

Consider first linear actions, and suppose for a curve $\mathbf{u}(s) \subset \mathbb{R}^n$ that $g \cdot \mathbf{u} = g\mathbf{u}$, and that the parameter s is invariant. Then if

$$\mathbf{u}_i = \frac{\mathrm{d}^i}{\mathrm{d}s^i}\mathbf{u}(s)$$

we have that

$$g \cdot \mathbf{u}_i = g\mathbf{u}_i.$$

Let $\mathbf{u}^{(N)}$ denote the derivatives of $\mathbf{u}$ up to order N, where N is such that G acts freely and regularly on some open subset of smooth curves and their derivatives up to order N. By definition, if $\rho = \rho(\mathbf{u}^{(N)})$ is a right frame, taking the space of smooth curves in $\mathbb{R}^n$ to G, then

$$\iota(\mathbf{u}_k) = \rho\mathbf{u}_k,$$

and for any $g \in G$ we have

$$\iota(\mathbf{u}_k) = \rho g^{-1} g \mathbf{u}_k. \tag{4.17}$$

Differentiating equation (4.17) with respect to the curve parameter s and setting $g = \rho$, we obtain

$$\frac{\mathrm{d}}{\mathrm{d}s}\iota(\mathbf{u}_k) = \iota(\mathbf{u}_{k+1}) + \rho_s \rho^{-1}\iota(\mathbf{u}_k). \tag{4.18}$$

Thus the error term in the invariant differentiation of $\iota(\mathbf{u}_k)$ is precisely $\rho_s \rho^{-1}\iota(\mathbf{u}_k)$. Comparing equation (4.18) with equations (4.12) and (4.13), and noting that the invariantised infinitesimal of the action on $\mathbf{u}_k$ is $\iota(\mathbf{u}_k)$, we see that $\mathbf{K}$ is indeed the invariantised Jacobian of the frame, not in coordinate form, but in matrix form.

The solution of the following exercise is discussed in detail in Lemma 5.2.1.

Exercise 4.5.12 Since the action is left and the frame is right, we have $\rho(g \cdot \mathbf{u}^{(N)}) = \rho(\mathbf{u}^{(N)})g^{-1}$. Hence show that the components of $\rho_s \rho^{-1}$ are invariants. Further, show $\rho_s \rho^{-1} \in T_e G = \mathfrak{g}$.

Applications of equation (4.18) are detailed in Section 5.5, where the affine case is discussed.

4.6 *Recursive construction of frames

There are several cases where a frame can be defined recursively, that is, where the frame can be written as a product of frames for lower dimensional group actions. This topic was explored by Kogan (2000a, 2000b). In this section we consider first the simplest case, and then generalise in various ways.

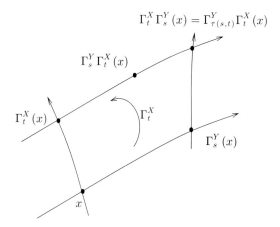

$$\Gamma_t^X \Gamma_s^Y (x) = \Gamma_{\tau(s,t)}^Y \Gamma_t^X (x)$$

$$\Gamma_s^Y \Gamma_t^X (x)$$

$$\Gamma_t^X (x)$$

$$\Gamma_t^X$$

$$\Gamma_s^Y (x)$$

$$x$$

Figure 4.10 If $[X, Y] = \lambda Y$, then the flow of the vector field X maps flowlines of Y to flowlines of Y.

Consider the two parameter group action

$$\widetilde{x} = x, \qquad \widetilde{u} = \exp(\epsilon)u + k \qquad (4.19)$$

of scaling and translation. The infinitesimal action is

$$\mathbf{v}_\epsilon = u \partial_u, \qquad \mathbf{v}_k = \partial_u$$

and

$$[\mathbf{v}_\epsilon, \mathbf{v}_k] = -\mathbf{v}_k.$$

This is the simplest example of a *solvable* group, which we define below and which is one of the main examples where a frame can be defined recursively. Prolonging the action to $(x, u, u_x, u_{xx}, \dots)$-space, the group orbits are two dimensional except along the plane $u = u_x = u_{xx} = \cdots = 0$. By Exercise 3.1.17, with $Y = \mathbf{v}_k$ and $X = \mathbf{v}_\epsilon$, we have

$$\Gamma_s^X \Gamma_{\exp(-s)t}^Y = \Gamma_t^Y \Gamma_s^X. \qquad (4.20)$$

This implies that the flow map Γ_t^X of X maps the flowline of Y through x to the flowline of Y through $\Gamma_t^X(x)$. For this to be true, we need $\Gamma_t^X \Gamma_s^Y = \Gamma_{\tau(s,t)}^Y \Gamma_t^X$, see Figure 4.10. And indeed we have $\tau(s, t) = \exp(t)s$. This can be verified in this simple case by noting

$$\Gamma_t^X \Gamma_s^Y (u) = \Gamma_t^X (u + s) = \exp(t)(u + s) = \Gamma_{\exp(t)s}^Y \Gamma_t^X (u)$$

which is equivalent to equation (4.20).

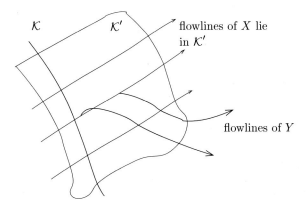

Figure 4.11 The cross section $\mathcal{K}'$ comprises flowlines of X; the flow of X maps the set of flowlines of Y to itself.

Let the transformation group $\mathcal{T}$ be that generated by the flows of both X and Y, that $[X, Y] = \lambda Y$, and that X and Y are linearly independent (nonzero). Suppose the action of $\mathcal{T}$ is free and regular. Then we can take a cross section $\mathcal{K}$ for $\mathcal{T}$ which defines a frame ρ. We will show that this can be written as a product of frames for the actions given by the flows of X and Y alone.

The first step is to use the flow of the vector field X to create a cross section

$$\mathcal{K}' = \cup_t \Gamma_t^X(\mathcal{K})$$

for the flow of the vector field Y; consider Figure 4.11.

We first prove that we can define a frame for $\mathcal{T}$ recursively using $\mathcal{K}'$, and then generalise to other intermediate cross sections.

By the Frobenius Theorem, the orbit $\mathcal{O}(z)$ of $\mathcal{T}$ through any one point z where the action is regular is a two dimensional surface, and the intersection of that surface with $\mathcal{K}$ will be a single point, which we will call k; see Figure 4.12.

Assume the action is a left action. Then a right frame for $\mathcal{T}$ for points on $\mathcal{O}(z)$ is a map $\rho : \mathcal{O}(z) \to \mathcal{T}$ such that $\rho(g * z) = \rho(z)g^{-1}$. Since $\mathcal{K}'$ is a cross section for the Y action, there is a Γ^Y-equivariant map $\rho_Y : \mathcal{O}(z) \to \{\Gamma_t^Y \mid t\}$ such that $k' = \rho_Y(z) * z \in \mathcal{K}'$. Then since $\mathcal{K}'$ is an orbit of X there is a unique Γ_s^X such that $\Gamma_s^X(k') = k$. Define $\rho_X(k') = \Gamma_s^X$. We thus have

$$\rho(z) = \rho_X(k') \cdot \rho_Y(z), \qquad k' = \rho_Y(z) * z. \tag{4.21}$$

Reversing the process just described, we have that given a frame ρ_Y for the action of the flow of Y relative to the cross section $\mathcal{K}'$, and a frame for the flow

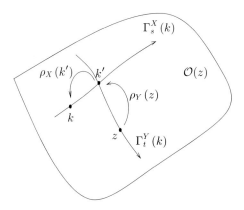

Figure 4.12 Defining a frame recursively. On the orbit $\mathcal{O}(z)$ of z, we have $\{k\} = \mathcal{K} \cap \mathcal{O}(z)$ and $\rho(z) = \rho_X(k')\rho_Y(z)$, where $k' = \rho_Y(z) * z$.

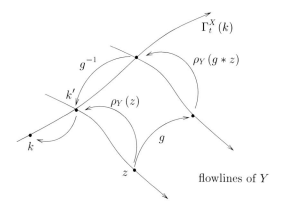

Figure 4.13 Diagram for the proof of equivariance in the case $g = \Gamma_s^X$, some s.

of X restricted to the cross section $\mathcal{K}'$, we may use equation (4.21) to define a frame for $\mathcal{T}$. (We remove the need for $\mathcal{K}'$ to consist of flowlines of X below.)

Lemma 4.6.1 *The recursively defined right frame (4.21) is equivariant.*

Proof We assume the action is left. To prove that ρ is equivariant, that is, $\rho(g * z) = \rho(z)g^{-1}$, there are two cases to consider. The first is $g = \Gamma_s^Y$, which follows from

$$\rho_Y(g * z) = \rho_Y(z) \cdot g^{-1}, \quad \text{and} \quad \rho_Y(g * z) * (g * z) = \rho_Y(z) * z.$$

For the second case, $g = \Gamma_s^X$, consider Figure 4.13. Even though $\rho_Y(g * z) \neq \rho_Y(z)$ in general, indeed equality holds only if the X and Y flows commute, we

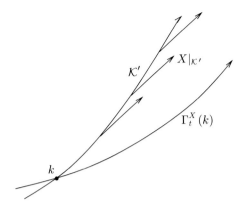

Figure 4.14 Generalising the intermediate cross section. If the projection of $X|_{\mathcal{K}'}$ to $T\mathcal{K}'$ is non-zero, the induced flow on $\mathcal{K}'$ can be used to define ρ_X.

still have $g^{-1}\rho_Y(g*z) = \rho_Y(z)g^{-1}$ since both these group elements take $g*z$ to k' and the action is free. We then have

$$\rho(g*z) = \rho_X(k')g^{-1}\rho_Y(g*z) = \rho_X(k')\rho_Y(z)g^{-1}$$

as required. $\square$

Finally, we remove the need for the intermediate cross section to be composed of flowlines of X. Consider Figure 4.14, where we have depicted the vector field X restricted to the proposed new intermediate cross section $\mathcal{K}'$. Provided the projection of $X|_{\mathcal{K}'}$ to the tangent space $T\mathcal{K}'$ of $\mathcal{K}'$ is non-zero, there will be an induced flow on $\mathcal{K}'$ that can be used to define ρ_X on $\mathcal{K}'$. We leave it to the reader to check that the frame thus defined satisfies the appropriate equivariance condition.

From the discussion in Section 4.4, we know that the cross section for the flow of Y has coordinates that are invariants of that flow. Thus, the action of Γ^X induces an action on those invariants. Consider again the simple scaling and translation action, equation (4.19). By the theory we have developed, there will be a scaling action on the translation invariants. These are u_x, u_{xx}, $u_{xxx}, \ldots$ and the induced scaling action is

$$\widetilde{u_{x\cdots x}} = \exp(\epsilon)u_{x\cdots x}.$$

The frame ρ_X is then a frame for this induced scaling action.

Exercise 4.6.2 Show that translation does *not* induce an action on the scaling invariants of the action (4.19). Thus show that a recursive frame *cannot* be defined for the action (4.19) by first using a frame for the scaling action followed by one for the translation action.

Exercise 4.6.3 Suppose that the action of $\mathcal{T}$ is right, not left. How does this affect the formula for the recursively defined frame?

Exercise 4.6.4 Consider the two parameter Lie group with infinitesimal vector fields

$$\mathbf{v}_a = x\partial_x - 2\partial_y$$
$$\mathbf{v}_b = x\log x\,\partial_x - 2(1 + \log x)\partial_y.$$

These arise as the infinitesimals of the symmetry group of the ordinary differential equation

$$y_{xx} + \frac{1}{x}y_x + e^y = 0 \tag{4.22}$$

see Cantwell (2002), Example 8.2. Show that

$$[\mathbf{v}_a, \mathbf{v}_b] = \mathbf{v}_a.$$

Show that one parametrisation of the group action is

$$\widetilde{x} = ax^b, \qquad \widetilde{y} = y - 2\log(ab) - 2(b-1)\log x$$

with identity element $(b, a) = (1, 1)$, and deduce that the group is $G = \mathbb{R}^+ \ltimes \mathbb{R}^+$ (or $(\mathbb{C} \setminus \{0\}) \ltimes (\mathbb{C} \setminus \{0\})$), depending on whether x and y are real or complex variables) with product

$$(\beta, \alpha) \cdot (b, a) = (\beta b, a\alpha^b).$$

Hint: the action on x and y will be a right action while the group product refers to a left action of G on itself. Show that a matrix representation of this group is given by

$$(b, a) \rightarrow \begin{pmatrix} b & \log a \\ 0 & 1 \end{pmatrix}.$$

Show that

$$\mathcal{D} = x\frac{\mathrm{d}}{\mathrm{d}x}$$

is an invariant differential operator for the flow of $Y = \mathbf{v}_a$, and that the differential invariants of the flow of Y are functions of

$$x^2\exp y, \qquad x^n\frac{\mathrm{d}^n}{\mathrm{d}x^n}y, \qquad n = 1, 2, \ldots$$

Show that the flow of $X = \mathbf{v}_b$ induces an action on the invariants of the flow of $Y = \mathbf{v}_a$. The normalisation equations $\widetilde{x} = 1$ and $\widetilde{y} = 0$ yield the frame, in parameter form,

$$b = x \exp(y/2), \qquad a = \exp(-x \log x \exp(y/2)).$$

Show that the frame can be obtained recursively and write the frame in matrix form as a product. Note: we will use moving frames to integrate equation (4.22) in Chapter 6.

There are several ways in which the above discussion can be generalised. The first is to the case where the transformation group T can be expressed as the product of two subgroups, $T = \mathcal{G}\mathcal{H}$ such that if $\mathfrak{g}$ is the Lie algebra of $\mathcal{G}$ and $\mathfrak{h}$ is the Lie algebra of $\mathcal{H}$ then

$$[x, y] \in \mathfrak{h}, \text{ for all } x \in \mathfrak{g}, y \in \mathfrak{h}. \tag{4.23}$$

Then the action of $\mathcal{G}$ maps orbits of $\mathcal{H}$ to orbits of $\mathcal{H}$. One can then define a right frame for T as $\rho = \rho_{\mathcal{G}}\rho_{\mathcal{H}}$.

Remark 4.6.5 If equation (4.23) holds, we say that $\mathfrak{h}$ is an ideal of the Lie algebra of T.

Exercise 4.6.6 In Section 1.2.1 we defined the semi-direct product of two Lie groups. Consider the special affine group $SA(2) = SL(2) \ltimes \mathbb{R}^2$ given in Example 1.2.19. Use the standard representation given in equation (1.21) to write down a basis for $\mathfrak{sa}(2)$. Prove the subalgebra corresponding to $\mathbb{R}^2$ is an ideal of $\mathfrak{sa}(2)$. Consider the space (x, u, v). Assume x is invariant, taking the standard linear action

$$\begin{pmatrix} \widetilde{u} \\ \widetilde{v} \end{pmatrix} = \begin{pmatrix} a & b \\ c & d \end{pmatrix} \begin{pmatrix} u \\ v \end{pmatrix} + \begin{pmatrix} k_1 \\ k_2 \end{pmatrix} \tag{4.24}$$

where $ad - bc = 1$, setting $u = u(x)$ and $v = v(x)$ and prolonging the action to $(u, v, u_x, v_x, u_{xx}, v_{xx})$-space, so the action becomes free and regular, show that the $SL(2)$ part of the group takes the set of orbits of the translation group $\mathbb{R}^2$ to itself, and further induces an action on the translation invariants. Obtain the frame as a product and show it is equivariant. Hint: the calculations are much easier if the inverse action is taken, see Example 4.1.6; in this case $\rho = \rho_{\mathbb{R}^2}\rho_{SL(2)}$ since the action is now a right action. Use the standard representation in equation (1.21) to write the frame ρ as a matrix product.

A second generalisation is to the case where the group is *solvable*.

Definition 4.6.7 A Lie group G is said to be *solvable* if there is a basis $\mathbf{v}_1$, $\mathbf{v}_2, \ldots, \mathbf{v}_r$ for its Lie algebra $\mathfrak{g}$ such that

$$[\mathbf{v}_i, \mathbf{v}_j] = \sum_{k \geq \max\{i,j\}} c_{ijk} \mathbf{v}_k. \tag{4.25}$$

We also say the Lie algebra $\mathfrak{g}$ is solvable.

Since $\langle \mathbf{v}_k, \mathbf{v}_{k+1}, \ldots, \mathbf{v}_n \rangle$ generates an ideal of $\langle \mathbf{v}_{k-1}, \mathbf{v}_k, \mathbf{v}_{k+1}, \ldots, \mathbf{v}_n \rangle$ for each $k = 2, \ldots, n$, a frame for a free and regular action of G can be written as an n-fold product.

The standard example of a solvable group is that of invertible upper triangular matrices. In fact, any solvable group can be represented by a matrix group whose elements are upper triangular.

Exercise 4.6.8 Show that the Lie group of real invertible 3×3 upper triangular matrices has Lie algebra

$$\mathfrak{g} = \left\{ \begin{pmatrix} a_1 & a_2 & a_3 \\ 0 & a_4 & a_5 \\ 0 & 0 & a_6 \end{pmatrix} \mid a_i \in \mathbb{R} \right\}.$$

Let the basis elements $\mathbf{w}_i$ be defined by

$$\begin{pmatrix} a_1 & a_2 & a_3 \\ 0 & a_4 & a_5 \\ 0 & 0 & a_6 \end{pmatrix} = \sum_i a_i \mathbf{w}_i.$$

Find a change of basis such that the new basis satisfies equation (4.25).

Exercise 4.6.9 Consider the three vector fields on (x, y)-space,

$$\begin{aligned} \mathbf{w}_1 &= (y - xy^2)\partial_y \\ \mathbf{w}_2 &= \tfrac{1}{2}x^2 y^2 \partial_y \\ \mathbf{w}_3 &= y^2 \partial_y. \end{aligned} \tag{4.26}$$

Note x is invariant under the action induced by the flows of the vector fields. Show these generate a solvable Lie algebra by finding a change of basis, to $\mathbf{v}_i$, $i = 1, 2, 3$ that satisfies equation (4.25). Obtain the action on y generated by the $\mathbf{v}_i$. To make the action free and regular on some domain, prolong it sufficiently, to $(x, y, y_x, y_{xx}, y_{xxx}, \ldots)$-space, and obtain a frame for the action. Show that the group underlying the action can be represented as the subgroup of the set of invertible 3×3 upper triangular matrices, given by

$$\left\{ \begin{pmatrix} \exp(t) & s & r \\ 0 & 1 & 0 \\ 0 & 0 & 1 \end{pmatrix} \mid r, s, t \in \mathbb{R} \right\}.$$

Show that $\mathbf{v}_1$ and $\mathbf{v}_2$ induce a flow on the invariants of the flow of $\mathbf{v}_3$, and that $\mathbf{v}_1$ induces a flow on the invariants of the flows of both $\mathbf{v}_2$ and $\mathbf{v}_3$. Hence show that a matrix representation of your frame for this action can be written as a triple product.

A classical theorem of the subject is that an ordinary differential equation of order n with a solvable symmetry group of dimension n can be integrated by quadratures. A moving frame demonstration of this theorem will be given in Chapter 6, together with a *far simpler* integration procedure. One of the joys of the moving frame approach is that the solvability of the symmetry group is much less important for success in integrating an ordinary differential equation than in the classical theory.

4.7 *Joint invariants

This section concerns invariants for actions on sets of points, sets of curves and so on, in other words, for the product action. Invariants of product actions are called N-point invariants and were defined in Definition 1.3.6. Moving frames were used to find N-point invariants for the Euclidean action in Example 4.4.4. Differential invariants for such actions involving more than one dependent variable are called *joint invariants*. They are obtained by the same process as described earlier, by evaluating the transformed variables on a chosen frame.

Example 4.7.1 Consider the action on smooth curves in $(u_1, u_2, \dots)$-space, given by

$$\widetilde{s} = s, \quad \widetilde{u}_i = \exp(a)u_i + b, \quad i = 1, 2, \dots$$

where $a, b \in \mathbb{R}$ and where the action on each dependent variable is the same. For the normalisation equations

$$\widetilde{u}_1 = 0, \qquad \widetilde{u_{1,s}} = 1$$

the frame is $\exp(a) = 1/u_{1,s}$, $b = -u_1/u_{1,s}$ and the joint invariants are

$$I^k = \frac{u_k - u_1}{u_{1,s}}, \qquad I_1^k = \frac{u_{k,s}}{u_{1,s}}, \qquad I_{11}^k = \frac{u_{k,ss}}{u_{1,s}}$$

and so on. For the normalisation equations $\widetilde{u}_1 = 0$, $\widetilde{u}_2 = 1$, the frame is $\exp(a) = 1/(u_2 - u_1)$, $b = -u_1/(u_2 - u_1)$ and the joint invariants are

$$I^k = \frac{u_k - u_1}{u_2 - u_1}, \qquad I_1^k = \frac{u_{k,s}}{u_2 - u_1}, \qquad I_{11}^k = \frac{u_{k,ss}}{u_2 - u_1}$$

and so on.

It can be seen that the set of joint invariants includes the N-point invariants.

In this next well-known example, we show that the solution of Liouville's equation is a joint invariant of the standard $SL(2)$ action.[†]

Example 4.7.2 Liouville's equation is

$$u_{xt} = \alpha \exp(\beta u),$$

where α and β are non-zero constants that can be changed by scaling u and x. We consider the product action on (f, g) by $SL(2)$,

$$(\widetilde{f}, \widetilde{g}) = \left(\frac{af + b}{cf + d}, \frac{ag + b}{cg + d} \right), \qquad ad - bc = 1,$$

$\widetilde{x} = x, \widetilde{t} = t$, and where the key trick is to set $f = f(x)$, $g = g(t)$. The solution of Liouville's equation is then the lowest order joint invariant. The restricted dependences of f and g yield, for any frame,

$$\begin{cases} 0 = I_K^f, & K = 11 \cdots 1 \underbrace{22 \cdots 2}_{n} \quad n > 0 \\ 0 = I_K^g, & K = \underbrace{11 \cdots 1}_{m} 22 \cdots 2 \quad m > 0. \end{cases}$$

If the normalisation equations are

$$\widetilde{f} = 0, \qquad \widetilde{f}_x = 1, \qquad \widetilde{g} = 1$$

then the invariant derivatives of the lowest order invariants are

$$\frac{\partial}{\partial t} I_{11}^f = -2 I_2^g$$

$$\frac{\partial}{\partial x} I_2^g = \left(I_{11}^f + 2 \right) I_2^g.$$

Hence the equation satisfied by I_2^g is

$$\frac{\partial^2}{\partial x \partial t} \log I_2^g = -2 I_2^g.$$

[†] This example was calculated after listening to a lecture by Ian Anderson given in Sommerøy on the problem of deciding when a partial differential equation was Darboux integrable.

Setting $u = \log I_2^g$ yields a solution of Liouville's equation with $\alpha = -2$ and $\beta = 1$. Calculating the invariant I_2^g explicitly, by solving for the frame, yields the well-known solution.

We leave it to the reader to find joint invariants of other actions, and the partial differential equations satisfied by the joint invariants. The inverse problem, that of deciding when a partial differential equation can be solved in terms of the joint invariants of some (unspecified) Lie group action, is related to the problem of deciding when a differential equation is Darboux integrable.

5
On syzygies and curvature matrices

In Chapter 4, Section 4.5, we introduced the moving frame for a prolonged action, and in particular considered the distinguished set of invariant differential operators $\{\mathcal{D}_j\}$ that the frame yields, as well as the formulae for the differentiation of the symbolic invariants, I_K^α.

The existence of the correction terms N_{ij} and M_{Kj}^α in the formulae

$$\mathcal{D}_j J_i = \delta_{ij} + N_{ij}, \qquad \mathcal{D}_j I_K^\alpha = I_{Kj}^\alpha + M_{Kj}^\alpha$$

has profound consequences for the study of invariant differential systems. In this chapter, we consider the main features of the 'landscape' of the invariant calculus. In particular, we consider finite sets of *generators* of the differential algebra of invariants, and the functional and differential relations, called *syzygies*, that they satisfy.

For simple normalisation equations, naive 'lattice diagrams' can be drawn that show the location of generating invariants and syzygies, and we show how to draw these. The theorems guaranteeing explicit finite sets of generating sets of generators and syzygies in terms of the normalisation equations, obtained by Hubert (2009a), are subtler to state and prove than were originally thought; the statements in the original Fels and Olver (1998, 1999) papers were proved later to hold only for so-called minimal frames (Definition 5.1.4). Simpler to obtain are the generators found as components of the correction matrix **K**, given in Theorem 4.5.6, and syzygies found as the components of the 'Maurer–Cartan equation', Proposition 5.2.8, satisfied by these generators. For applications we consider, these syzygies suffice; at the time of writing, it is not known whether these form a complete set of generators for the set of syzygies or not.

The second part of the chapter is devoted to the *curvature matrices*

$$\mathcal{Q}_i = \mathcal{D}_i \varrho \cdot \varrho^{-1}$$

where ϱ is a matrix representation of the frame. These matrices provide a computational link between a differential system and its invariantisation. For example, they allow Lie group based numerical integrators to be used to integrate ordinary differential equations numerically, as explained in Chapter 6. The major result we prove is that the matrices Q_i can be calculated for any given representation of the Lie algebra of the group, using only the infinitesimals of the group action and the normalisation equations. In particular, it is not necessary to solve for the frame.

The curvature matrices are well known in Differential Geometry. We examine in depth the most famous example, the Serret–Frenet frame for the action of the special Euclidean group on curves in $\mathbb{R}^3$, and show how the classical results appear in the symbolic invariant calculus notation. We then consider some results for linear actions on curves in $\mathbb{R}^n$ induced by a matrix Lie group.

Notation   Throughout this chapter, we continue with the notation used in earlier chapters. We will assume that the Lie group G acts on the space with local coordinates

$$(x_1, x_2, \ldots, x_p, u^1, u^2, \ldots, u^q).$$

The x_k are the independent variables, $u^\alpha = u^\alpha(x_1, x_2, \ldots, x_p)$ and the action is prolonged to the jet space with local coordinates

$$(\ldots, x_k, \ldots, u^\alpha, \ldots, u_K^\alpha, \ldots)$$

where K denotes the multi-index of differentiation with respect to the x_k and where the '...' will be left implicit. The normalisation equations are denoted as

$$\psi_j(x_k, u^\alpha, u_K^\alpha) = 0, \quad j = 1, \ldots, r = \dim(G).$$

The frame is denoted as ρ. The invariantisation map is given as

$$\iota(F(x_k, u^\alpha, u_K^\alpha)) = F(g \cdot x_k, g \cdot u^\alpha, (g \cdot u_K^\alpha))|_{g=\rho} = F(J_k, I^\alpha, I_K^\alpha),$$

and the distinguished invariant differential operators are

$$\mathcal{D}_k = \widetilde{D}_k|_{g=\rho},$$

where $\widetilde{D}_k$ is defined in equation (1.35).

5.1 Computations with differential invariants

To best appreciate the major differences between the set of derivative terms $\{u_K^\alpha\}$ and the set of their corresponding invariants $\{I_K^\alpha\}$, we recall first some simple

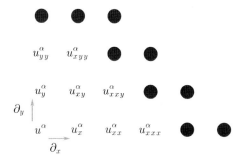

Figure 5.1 The 'differential structure' of $\{u_K^\alpha \mid |K| \geq 0\}$.

facts. For a given smooth function u, the set $\{u_K^\alpha\}$ can be obtained by acting with differentiation operators on the 'fundamental' set $\{u^\alpha\}$, $\alpha = 1, \ldots, q$ of dependent variables. Every derivative term u_K^α is generated by applying the derivative operators $\partial/\partial x_i$ the requisite number of times. As coordinates of a prolonged space, they are independent; there are no functional relations between them. Furthermore the only differential relations between the u_K^α are of the form $D^K u_{LM}^\alpha = D^M u_{LK}^\alpha$ for indices K, L and M, all of which are consequences of the fact that the operators commute. This situation is represented in Figure 5.1.

By contrast, we have the following.

(i) For fixed α, the set $\{I_K^\alpha \mid |K| \geq 0\}$ may not be generated by I^α under invariant differentiation. Indeed, I^α may be normalised to a constant.
    In general more than one generator will be needed to obtain the complete set, $\{I_K^\alpha\}$, under invariant differentiation.
(ii) The invariant differential operators may not commute.
(iii) There are functional relations between the invariants, given by the normalisation equations.
(iv) Non-trivial differential relations, known as *syzygies*, may exist.

Further, when computing with invariant derivatives symbolically, the following is also important.

(v) Invariant derivatives of the I_K^α may not be of polynomial type. If the normalisation equations are polynomials and the infinitesimals are rational, then the correction terms are rational. Their denominators consist of products of a finite number of factors which can be determined in advance.

We now look at these items in turn, beginning with the question of generators of the algebra of differential invariants. We know from Theorem 4.4.10 that any differential invariant can be written in terms of the J_i and the I_K^α.

Definition 5.1.1 We denote by $\mathcal{I}$ the algebra generated by the set

$$\{J_i, I_K^\alpha \mid i = 1, \ldots, p \,; \alpha = 1, \ldots, q; \ |K| \geq 0\}.$$

Since any function of invariants is invariant, the set $\mathcal{I}$ consists of invariants. Further, invariant differential operators take differential invariants to invariants. Since any invariant differential operator can be written in terms of the $\mathcal{D}_j$, $j = 1, \ldots, p$ and the elements of $\mathcal{I}$, then assuming the correction terms are rational, $\mathcal{I}$ is a differential algebra with respect to the $\mathcal{D}_j$.

Definition 5.1.2 We say that the finite subset $\mathcal{I}_{\text{gen}} \subset \mathcal{I}$ is *a set of generators* if any element $I \in \mathcal{I}$ can be written as a function of the elements of $\mathcal{I}_{\text{gen}}$ and a finite number of their invariant derivatives.

A classical theorem (Tresse, 1894) states that a finite set of generators exists. Since we have the formulae (4.11), a first result is that a set of generators is given by the (non-normalised) J_k and those I_K^α with $|K| \leq 1 + |\psi|$, where $|\psi|$ is the order of the normalisation equations.

Example 5.1.3 Consider the action of $SL(2) \ltimes \mathbb{R}^2$ on curves $x \mapsto (u(x), v(x)) \in \mathbb{R}^2$ given by

$$\begin{pmatrix} \widetilde{u} \\ \widetilde{v} \end{pmatrix} = \begin{pmatrix} a & b \\ c & d \end{pmatrix} \begin{pmatrix} u \\ v \end{pmatrix} + \begin{pmatrix} k_1 \\ k_2 \end{pmatrix}, \qquad ad - bc = 1$$

where $\widetilde{x} = x$ is the curve parameter. Since the single independent variable is invariant, we have $\mathcal{D} = \mathrm{d}/\mathrm{d}x$. Take the normalisation equations to be

$$\widetilde{u} = \widetilde{v} = 0, \qquad \widetilde{u}_x = \widetilde{v}_{xx} = 0, \qquad \widetilde{v}_x = 1,$$

so that $I^u = I^v = 0$, $I_1^u = I_{11}^v = 0$ and $I_1^v = 1$. Since there is only one independent variable, the diagram of invariants analogous to Figure 5.1 is also one dimensional; there is one diagram for each dependent variable. They look like

$$u: \qquad 0 \qquad 0 \qquad I_{11}^u \xrightarrow{\ \mathcal{D}\ } I_{111}^u \longrightarrow \bullet \quad \bullet \quad \bullet$$

$$v: \qquad 0 \qquad 1 \qquad 0 \qquad I_{111}^v \xrightarrow{\ \mathcal{D}\ } I_{1111}^v \longrightarrow \bullet \quad \bullet$$

The arrow in the diagram does not mean that $\mathcal{D}I_{11}^u = I_{111}^u$, but rather that I_{111}^u can be obtained in terms of $\mathcal{D}I_{11}^u$ and other invariants. From the normalisation equations, we see that the equations for the group parameters that define the frame involve at most second order quantities. Hence, since I_{11}^u is the second order quantity evaluated on the frame, it is a second order invariant. Similarly, both I_{111}^u and I_{111}^v are third order invariants. The table of infinitesimals is

$$
\begin{array}{c}
\\ k_1 \\ k_2 \\ a \\ b \\ c
\end{array}
\begin{array}{c}
\begin{array}{ccccc}
x & u & v & u_K & v_K
\end{array}\\
\left(\begin{array}{ccccc}
0 & 1 & 0 & 0 & 0 \\
0 & 0 & 1 & 0 & 0 \\
0 & u & -v & u_K & -v_K \\
0 & v & 0 & v_K & 0 \\
0 & 0 & u & 0 & u_K
\end{array}\right).
\end{array}
$$

The **K** matrix is

$$
\mathbf{K} = x \begin{array}{c}
\begin{array}{ccccc}
a & b & c & k_1 & k_2
\end{array}\\
\left(\begin{array}{ccccc}
0 & -I_{11}^u & -\dfrac{I_{111}^v}{I_{11}^u} & 0 & -1
\end{array}\right).
\end{array}
$$

Thus, we have the invariant differentiation formulae,

$$
\mathcal{D}I_{11}^u = I_{111}^u, \qquad \mathcal{D}I_{111}^u = I_{1111}^u - I_{11}^u I_{111}^v
$$

and

$$
\mathcal{D}I_{111}^v = I_{1111}^v - \frac{I_{111}^v}{I_{11}^u} I_{111}^u
$$

and so forth, and since the derivative of an nth order invariant is of order $n+1$, we see that I_{111}^v and I_{11}^u *generate* all the I_K^α by invariant differentiation.

Example 5.1.3 above is an example of a *minimal moving frame* (see Hubert 2009a, Section 4.2). We give a working definition here.

Definition 5.1.4 (Working definition) A *minimal frame* for a prolonged Lie group action is obtained by using normalisation equations that solve for as many group parameters as possible at every order of prolongation.

The next example contrasts the 'landscape of invariants' given by a minimal frame and a non-minimal frame for a simple group action.

Example 5.1.5 Consider the simple situation when the action on curves in the plane parametrised by $x \mapsto (x, u(x))$ is

$$
\widetilde{x} = \lambda x, \qquad \widetilde{u} = u.
$$

The lowest possible order normalisation equation is $\widetilde{x} = 1$, yielding the frame, $\lambda = 1/x$. The invariants are

$$
I^u = u, \qquad I_1^u = xu_x, \qquad \dots \qquad I_n^u = x^n \frac{\mathrm{d}^n u}{\mathrm{d}x^n}.
$$

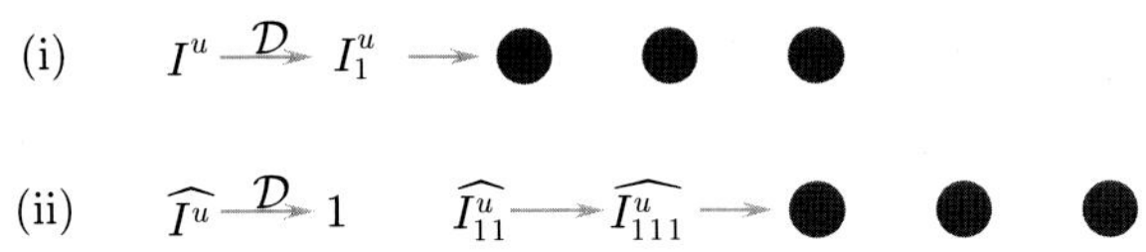

(i) $I^u \xrightarrow{\mathcal{D}} I_1^u \longrightarrow \bullet \quad \bullet \quad \bullet$

(ii) $\widehat{I^u} \xrightarrow{\mathcal{D}} 1 \quad \widehat{I_{11}^u} \longrightarrow \widehat{I_{111}^u} \longrightarrow \bullet \quad \bullet \quad \bullet$

Figure 5.2   Diagrams of invariants for Example 5.1.5.

where we have used I_n^u to denote $I_{1\ldots1}^u$ with n 1s in the index. Since the invariant differential operator is

$$\mathcal{D} = \frac{1}{x}\frac{d}{dx},$$

we have that I^u is the single generator. The diagram of invariants is given in Figure 5.2 (i).

But suppose that we take the normalisation equation to be $\widetilde{u}_x = 1$. The frame is now $\lambda = u_x$. We distinguish the system of invariants obtained with this second frame, from those of the previous frame, by a $\frown$ (a 'hat'). We have

$$\widehat{J} = xu_x, \qquad \widehat{I^u} = u, \qquad \widehat{I_n^u} = \frac{1}{u_x^n}\frac{d^n u}{dx^n}$$

and the invariant differential operator is now

$$\widehat{\mathcal{D}} = \frac{1}{u_x}\frac{d}{dx}.$$

The diagram of invariants is given in Figure 5.2 (ii). The diagram indicates that in addition to $\widehat{J}$, we need two additional generators, $\widehat{I^u}$ and $\widehat{I_{11}^u}$ (recall $\widehat{I_1^u} = 1$). But we also have two *syzygies*, namely

$$\widehat{\mathcal{D}}\widehat{J} = 1 + \widehat{I_{11}^u}\widehat{J}, \qquad \widehat{\mathcal{D}}\widehat{I^u} = 1.$$

It can be seen that a minimal set of generators is thus $\widehat{I^u}$ and $\widehat{J}$, that we cannot reduce the number of generators further, and that we have still the one relation on the generators.

It should not be assumed that invariants J_k play the role of the independent variables in the standard differential rings. The above Example 5.1.5 serves as a warning. Indeed, the invariant derivative of the invariant corresponding to the independent variable, $\widehat{J}$, is in terms of the second order invariant $\widehat{I_2^u}$.

We now look at some higher dimensional examples. The first example illustrates the fact that invariant differential operators may not commute.

Example 5.1.6 If the group is $SO(2) \ltimes \mathbb{R}^2$ acting on the independent variables as

$$\begin{pmatrix} \widetilde{x} \\ \widetilde{y} \end{pmatrix} = \begin{pmatrix} \cos(\theta) & \sin(\theta) \\ -\sin(\theta) & \cos(\theta) \end{pmatrix} \begin{pmatrix} x - a \\ y - b \end{pmatrix}$$

with the dependent variable $u = u(x, t)$ being an invariant, then a moving frame

$$\rho(x, y, u, u_x, u_y, \dots) = (\arctan(-u_x/u_y), (x, y))$$

is obtained from the normalisation equations $\widetilde{x} = 0$, $\widetilde{y} = 0$, $\widetilde{u}_x = 0$. Thus

$$J_1 = 0, \qquad J_2 = 0, \qquad I^u = u, \qquad I_1^u = 0$$

while the invariants corresponding to u_y, u_{xx} are

$$I_2^u = \sqrt{u_x^2 + u_y^2}$$

$$I_{11}^u = \frac{u_y^2 u_{xx} - 2u_x u_y u_{xy} + u_x^2 u_{yy}}{u_x^2 + u_y^2}.$$

The right hand sides of these can be checked independently to be invariants. Applying the Replacement Theorem, Theorem 4.4.9, verifies the left hand sides.

The two invariant differential operators obtained by evaluating $\partial/\partial\widetilde{x}$ and $\partial/\partial\widetilde{y}$ on the frame are

$$\mathcal{D}_1 = \frac{1}{\sqrt{u_x^2 + u_y^2}} \left(u_y \partial_x - u_x \partial_y\right)$$

$$\mathcal{D}_2 = \frac{1}{\sqrt{u_x^2 + u_y^2}} \left(u_x \partial_x + u_y \partial_y\right).$$

These operators do not commute. In fact,

$$[\mathcal{D}_1, \mathcal{D}_2] = \frac{I_{11}^u}{I_2^u}\mathcal{D}_1 + \frac{I_{12}^u}{I_2^u}\mathcal{D}_2.$$

Despite the fact that $I_1^u = 0$ and thus $\mathcal{D}_1 I_1^u = 0$, the invariant I_{11}^u is *not* zero. As an example of the correction terms arising with invariant differentiation, we have that

$$\mathcal{D}_1 I_{11}^u = I_{111}^u - 2\frac{I_{11}^u I_{12}^u}{I_2^u}$$

which is not of polynomial type. In Figure 5.3, we show the 'differential structure' of the set of invariants. In this example there are *two* generating invariants, namely I^u and I_{11}^u: it is not possible to obtain I_{11}^u from I^u using invariant differentiation; note both I_1^u and $\mathcal{D}_1^u$ are zero.

Because the formulae for the correction terms are so complex, it seems a difficult question to say in advance just what the smallest set of generators and

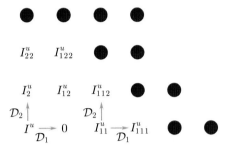

Figure 5.3 The 'differential structure' of $\{I_K^\alpha \mid |K| \geq 0\}$ for Example 5.1.6.

relations are, for a given frame. The difficulty is compounded when one has normalisation equations for cross sections that are not coordinate planes, and in particular that are non-linear.

Drawing lattice diagrams of invariants and calculating symbolic invariant derivatives is a good way to begin to investigate which of the I_K^α may be generators. Two distinguished sets are the following.

Definition 5.1.7 We denote by $\mathcal{I}^0$ the set of *zeroth invariants*,

$$\mathcal{I}^0 = \{\iota(x_j) = J_j, \iota(u^\alpha) = I^\alpha \mid j = 1, \ldots, p, \ \alpha = 1, \ldots, q\}.$$

Further, we denote by $\mathcal{E}$ the set of *edge invariants*,

$$\mathcal{E} = \left\{\iota\left(\frac{\partial \psi_k}{\partial x_j}\right) \mid k = 1, \ldots, r, \ j = 1, \ldots, p\right\}.$$

If the normalisation equations are for a cross section that is a coordinate plane, then the edge invariants can be located easily on the lattice diagram of invariants; if a normalisation equation is $u_K^{\widetilde{\alpha}} = c$ for some constant c, then its corresponding edge invariants are I_{Kj}^α, for $j = 1, \ldots, p$. For minimal frames, the edge invariants give an easily visualised generating set of differential invariants. The following result is proved for example in Hubert (2009a), Theorem 4.2.

Theorem 5.1.8 *If the normalisation equations $\psi_k = 0$, $k = 1, \ldots, r$, give a minimal frame, then $\mathcal{I}^0 \cup \mathcal{E}$ form a generating set of differential invariants.*

Theorem 5.1.8 is false if the frame is not minimal. However, many useful frames are not minimal. Hubert proves a generalisation of Theorem 5.1.8 which removes the requirement of minimality. The statement of the result uses concepts from the theory of over determined differential systems, and the proof uses results from non-commutative differential algebra. The result is important not least because it leads to a generating set of syzygies, the subject of Section 5.1.1. The next theorem, Theorem 5.1.9, also due to Hubert (Hubert, 2009b, Theorem 4.2) is useful in applications.

Theorem 5.1.9 *Suppose the normalisation equations* $\psi_k = 0$, $k = 1, \ldots, r$, *yield a frame for a regular free action on some open set of the prolonged space with coordinates* $(x_j, u^\alpha, u_K^\alpha)$. *Then the components of the correction matrix* **K**, *given in Theorem 4.5.6, together with* $\mathcal{I}^0$, *given in Definition 5.1.7, form a generating set of differential invariants.*

Theorem 5.1.9 means that once we have calculated the correction matrix **K**, which we need to do in order to use the symbolic differentiation formulae, then we have already calculated a relatively small set of generating differential invariants.

5.1.1 Syzygies

We now turn to the general question of relations and syzygies. In the previous section we noted that the set of invariants $\mathcal{I}$ is an algebra.

Definition 5.1.10 A *syzygy* S on the set $\mathcal{I}$ is a function of finitely many elements of $\mathcal{I}$ that is identically zero when written in terms of the coordinates $(x_k, u^\alpha, u_K^\alpha)$. The set of syzygies of $\mathcal{I}$ is denoted $\mathcal{S}(\mathcal{I})$.

Syzygies are often written as an *identity* or equation of the form

$$S(J_k, I^\alpha, I_K^\alpha) = 0,$$

where S has a finite number of arguments. In terms of a finite set of generators $\{\sigma_1, \ldots, \sigma_N\}$ of $\mathcal{I}$, a syzygy is an identity of the form

$$S(\sigma_k, \mathcal{D}^K \sigma_k) = 0$$

with a finite number of invariant derivatives of the σ_k appearing as arguments of S.

There are two distinguished sets of syzygies.

(i) The first is the set of invariantised normalisation equations. Since the frame ρ solves the equations $\psi(\rho \cdot x_k, \rho \cdot u^\alpha, \rho \cdot u_K^\alpha) = 0$, then by construction, $\psi(J_k, I^\alpha, I_K^\alpha) = 0$. If the dimension of G is r, we will have r normalisation equations which solve for the r parameters and hence $r = \dim(G)$ of these relations.

(ii) The second distinguished set is the infinite set of symbolic invariant differentiation formulae,

$$\mathcal{D}_j J_i = \delta_{ij} + N_{ij}, \qquad \mathcal{D}_j I_K^\alpha = I_{Kj}^\alpha + M_{Kj}^\alpha,$$

where δ_{ij} is the Kronecker delta.

Since functions of syzygies are again syzygies, and invariant derivatives of syzygies yield syzygies, the set $\mathcal{S}(\mathcal{I})$ is a differential algebra. Since we can multiply a syzygy by an invariant and still obtain a syzygy, we say that $\mathcal{S}(\mathcal{I})$ is a module over $\mathcal{I}$.

As with $\mathcal{I}$, we seek a small finite set of generators of $\mathcal{S}(\mathcal{I})$. A complete set of generators of the syzygy module for edge invariants (Definition 5.1.7) is given by Hubert (2009a), Theorem 5.14. This theorem requires concepts from the theory of over determined differential systems and non-commutative differential algebra to state and prove, so we concentrate instead on a discussion of the main types of syzygy and examples.

Two interesting kinds of syzygy are the following.

(i) If I_{Ki}^α is a normalised invariant but I_K^α is not, then in the equation $\mathcal{D}_i I_K^\alpha = I_{Ki}^\alpha + M_{Ki}^\alpha$, the term I_{Ki}^α must cancel. In terms of the lattice diagram, these syzygies arise when an invariant maps into a 'hole' in the diagram of invariants, as in Figure 5.2 (ii) where $\mathcal{D}\widehat{I^u} = 1$, or in Figure 5.3 where $\mathcal{D}_1 I = 0$.

(ii) If $p > 1$, another interesting kind of syzygy is possible. Let I_J^α, I_L^α be two (generating) differential invariants, and let indexes K, M be such that $JK = LM$. In the two equations for $\mathcal{D}_K I_J^\alpha$ and $\mathcal{D}_M I_L^\alpha$, the term $I_{JK}^\alpha = I_{LM}^\alpha$ must cancel, and then

$$\mathcal{D}_K I_J^\alpha - \mathcal{D}_M I_L^\alpha = M_{JK}^\alpha - M_{LM}^\alpha \tag{5.1}$$

is a syzygy.

Example 5.1.6 continued Consider Figure 5.4. There are two paths of differentiation from a generator to I_{112}^u. Taking the two symbolic differentiation formulae from I^u and I_{11}^u to I_{112}^u and cancelling the I_{112}^u term yields

$$\mathcal{D}_1^2 \mathcal{D}_2 I^u - \mathcal{D}_2 I_{11}^u = \frac{1}{I_2}\left(\left(I_{11}^u\right)^2 + 2\left(I_{12}^u\right)^2 - I_{11}^u I_{22}^u\right).$$

Since $\mathcal{D}_2 I^u = I_2^u$, $\mathcal{D}_1(I_2^u) = I_{12}^u$ and $\mathcal{D}_2(I_2^u) = I_{22}^u$, the syzygy can be written in terms of the generators I^u and I_{11}^u in the form

$$\mathcal{D}_2(I^u)\left(\mathcal{D}_1^2 \mathcal{D}_2 I^u - \mathcal{D}_2 I_{11}^u\right) - \left(\left(I_{11}^u\right)^2 + 2\mathcal{D}_1 \mathcal{D}_2(I^u) - I_{11}^u \mathcal{D}_2^2(I^u)\right) = 0,$$

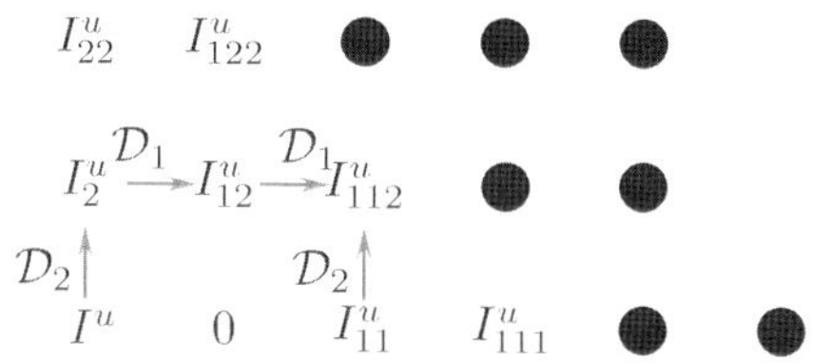

Figure 5.4 Visualising a syzygy for Example 5.1.6; two paths leading to I^u_{112} from generators yield a syzygy.

where recall that the order of differentiation matters. A different set of generators and their syzygies for this example is obtained as follows. The correction matrix **K** for this action is

$$
\mathbf{K} = \begin{array}{c} \\ x \\ y \end{array} \begin{array}{ccc} a & b & \theta \\ \left(\begin{array}{ccc} -1 & 0 & -\dfrac{I^u_{11}}{I^u_2} \\ 0 & -1 & -\dfrac{I^u_{12}}{I^u_2} \end{array} \right) \end{array}
$$

and hence by Theorem 5.1.9 we can take $\sigma_0 = I^u, \sigma_1 = I^u_{11}/I^u_2$ and $\sigma_2 = I^u_{12}/I^u_2$ to be a set of generators. The syzygies between σ_0 and σ_2, and between σ_1 and σ_2, are

$$
\mathcal{D}_2(\sigma_0)\mathcal{D}_1(\mathcal{D}_2(\sigma_0)) - \sigma_2 = 0
$$
$$
\mathcal{D}_1(\sigma_2) - \mathcal{D}_2(\sigma_1) + \sigma_1^2 + \sigma_2^2 = 0.
$$

5.2 Curvature matrices

We consider first the simplest one dimensional case, that of a curve $s \mapsto z(s) \in U$ where U is the domain of the moving frame $\rho : U \to G$. Then the image of the curve under the frame yields a curve in G, $s \mapsto G$, given by $s \mapsto \rho(z(s))$, see Figure 5.5.

Recall from Definition 1.2.1 that an n dimensional matrix representation $\mathcal{R}$ of a group G is a map $G \to GL(n, \mathbb{R})$ such that $\mathcal{R}(g)\mathcal{R}(h) = \mathcal{R}(gh)$. Note that this implies that if e is the identity element of G, then $\mathcal{R}(e)$ is the identity matrix and further $\mathcal{R}(g^{-1}) = \mathcal{R}(g)^{-1}$. Let $\varrho(z)$ denote the matrix $\varrho(z) = \mathcal{R}(\rho(z))$. Then $\varrho(s)$ is a path in the matrix representation space of G.

Lemma 5.2.1 *Given a group action on U, let $s \mapsto z(s) \in U$ be a path and assume that the parameter s is invariant under the group action, so that $\mathrm{d}/\mathrm{d}s$ is an invariant operator. Further, suppose the frame is given in matrix form, ϱ. If $\varrho : U \subset M \to G$ is a right frame for a left action, then*

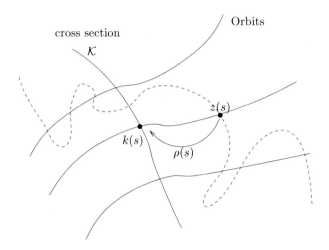

Figure 5.5 A right moving frame on a curve parametrised by s in U yields a path $\rho(s)$ in G.

(i) *the components of the matrix $\varrho_s \varrho^{-1}$ are invariants, and*

(ii) $\varrho_s \varrho^{-1} : U \to \mathfrak{g}$.

Proof The first follows, setting $z = z(s)$, from

$$\left(\frac{\mathrm{d}}{\mathrm{d}s}\varrho(g \cdot z)\right)\varrho(g \cdot z)^{-1} = \left(\frac{\mathrm{d}}{\mathrm{d}s}\varrho(z)g^{-1}\right)g\varrho(z)^{-1} = \left(\frac{\mathrm{d}}{\mathrm{d}s}\varrho(z)\right)\varrho(z)^{-1}$$

since the generic group element g is independent of s. To see the second, note that

$$\frac{\mathrm{d}}{\mathrm{d}s}\varrho(z(s)) \in T_{\varrho(z(s))}G$$

and hence right multiplication by $\varrho(z)^{-1}$ yields the result. (Recall for a matrix group, the tangent map induced by right multiplication is simply right multiplication.) □

We leave it to the reader to work out the corresponding invariant matrices for a left frame or a right action.

Lemma 5.2.1 generalises to images of ϱ on surfaces in U, with d/ds replaced by the invariant derivative operators $\mathcal{D}_i$.

Definition 5.2.2 We define the *curvature matrices*

$$\mathcal{Q}_i = (\mathcal{D}_i\varrho)\varrho^{-1}, \ i = 1, \ldots, p. \tag{5.2}$$

Exercise 5.2.3 Show that the components of the curvature matrices $\mathcal{Q}_i$ are invariant, and that $\mathcal{Q}_i : U \to \mathfrak{g}$.

Theorem 5.2.4, from Mansfield and van der Kamp (2006), provides a new significance for the correction matrix **K**; its rows are the coordinates of the curvature matrices, when expressed as a linear combination of the relevant basis of the Lie algebra.

If the coordinates near e in G are $(a_1, a_2, \ldots, a_r)$, then differentiating $\mathcal{R}(a_1, a_2, \ldots, a_r)$ with respect to a_i at e, we obtain

$$\mathfrak{a}_i = \frac{d\mathcal{R}(g)}{da_i}\bigg|_{g=e}, \qquad i = 1, \ldots, r \qquad (5.3)$$

which span a matrix representation of the Lie algebra $\mathfrak{g}$ of G induced from the representation $\mathcal{R}$ of G.

Theorem 5.2.4 *The curvature matrices $\mathcal{Q}_i$ can be constructed in a matrix representation of $\mathfrak{g}$, induced from a representation $\mathcal{R}$ of G, with basis $\{\mathfrak{a}_i\}$ defined in (5.3), using only the normalisation equations and the infinitesimal action. Indeed,*

$$\mathcal{Q}_i = \sum_j \mathbf{K}_{ij}\mathfrak{a}_j$$

*where **K** is the correction matrix given in Theorem 4.5.6.*

Remark 5.2.5 It is implicit in the statement of Theorem 5.2.4 that the parameters a_j used to calculate the matrices $\mathfrak{a}_j$ are the same as those used for the infinitesimal action and hence in the calculation of **K**. In practice, given a faithful representation, one checks that the Lie bracket multiplication table for the $-\mathfrak{a}_j$ (note the minus sign) is the same as that for the Lie bracket of the infinitesimal vectors. The change of sign is due to the change of parity in the action of G, which is left for the action of G on itself, but right for the action on functions on M.

Proof Choose $g \in G$ arbitrary with $\widetilde{z} = g * z$. On the one hand we have

$$\widetilde{D}_i \mathcal{R}(\rho(\widetilde{z}))\Big|_{g=\rho(z)} = \widetilde{D}_i (\mathcal{R}(\rho(z)))\mathcal{R}(g)^{-1}\Big|_{g=\rho(z)}$$
$$= \mathcal{Q}_i$$

and on the other hand

$$\widetilde{D}_i \mathcal{R}(\rho(\widetilde{z}))\Big|_{g=\rho(z)} = \sum_{j=1}^{r} \widetilde{D}_i \rho_j(\widetilde{z}) \frac{d\mathcal{R}(\rho(\widetilde{z}))}{d\rho_j(\widetilde{z})}\Big|_{g=\rho(z)}$$
$$= \sum_{j=1}^{r} \mathbf{K}_{ij}\mathfrak{a}_j$$

using the first expression for $\mathbf{K}_{ij}$ given in equation (4.13), and noting $\rho(\rho(z) * z) = e$. □

We demonstrate the theorem in two examples where the result can be checked since we know the frame explicitly.

Example 5.2.6 Consider the action of $SL(2)$ on $(x, u(x))$ space given by

$$\widetilde{x} = x, \qquad \widetilde{u} = \frac{au + b}{cu + d}, \qquad ad - bc = 1$$

prolong it to $(x, u, u_x, u_{xx}, \dots)$ space and take the frame $\widetilde{u} = 0$, $\widetilde{u}_x = 1$ and $\widetilde{u}_{xx} = 0$. Then the frame is

$$a = \frac{1}{\sqrt{u_x}}, \qquad b = -\frac{u}{\sqrt{u_x}}, \qquad c = \frac{u_{xx}}{2u_x^{3/2}}, \tag{5.4}$$

and

$$\begin{array}{ccc} a & b & c \end{array}$$
$$\mathbf{K} = x \begin{pmatrix} 0 & -1 & \frac{1}{2} I_{111}^u \end{pmatrix}.$$

In matrix form, we have

$$g(a, b, c) = \begin{pmatrix} a & b \\ c & (1 + bc)/a \end{pmatrix}$$

and thus

$$\mathfrak{a}_a = \begin{pmatrix} 1 & 0 \\ 0 & -1 \end{pmatrix}, \qquad \mathfrak{a}_b = \begin{pmatrix} 0 & 1 \\ 0 & 0 \end{pmatrix}, \qquad \mathfrak{a}_c = \begin{pmatrix} 0 & 0 \\ 1 & 0 \end{pmatrix}.$$

Inserting the frame parameters in equation (5.4) into $g(a, b, c)$ to obtain ϱ, we obtain directly that

$$\varrho_x \varrho^{-1} = \begin{pmatrix} 0 & -1 \\ \frac{1}{2}\left(\frac{u_{xxx}}{u_x} - \frac{3}{2}\frac{u_{xx}^2}{u_x^2}\right) & 0 \end{pmatrix}$$
$$= \begin{pmatrix} 0 & -1 \\ \frac{1}{2} I_{111}^u & 0 \end{pmatrix}.$$
$$= 0\mathfrak{a}_a + (-1)\mathfrak{a}_b + \frac{1}{2} I_{111}^u \mathfrak{a}_c$$
$$= \mathbf{K}_{xa}\mathfrak{a}_a + \mathbf{K}_{xb}\mathfrak{a}_b + \mathbf{K}_{xc}\mathfrak{a}_c.$$

Exercise 5.2.7 Verify the details in the following. Consider the action of $SL(2)$ on (x, u) space given by

$$\widetilde{x} = \frac{ax + b}{cx + d}, \qquad \widetilde{u} = 6c(cx + d) + (cx + d)^2 u, \qquad ad - bc = 1$$

prolong it to curves on this space, that is, to $(x(s), u(s), x_s, u_s, x_{ss}, \ldots)$ space and take the frame $\tilde{x} = 0$, $\tilde{u} = 0$ and $\tilde{x}_s = 1$. We have

$$
\mathsf{K} = x \quad \overset{\displaystyle a \quad\quad b \quad\quad c}{\left(-\tfrac{1}{2} I_{11}^x \quad -1 \quad -\tfrac{1}{6} I_1^u \right)}
$$

and

$$
a = \frac{1}{\sqrt{x_s}}, \qquad b = -\frac{x(s)}{\sqrt{x_s}}, \qquad c = -\frac{1}{6} u(s)\sqrt{x_s}.
$$

Using the same matrix representation of $SL(2)$ as in the previous example, inserting the values of a, b and c on the frame to yield ϱ and calculating $\varrho_s \varrho^{-1}$ both directly and using Theorem 5.2.4, yields

$$
\varrho_s \varrho^{-1} = \begin{pmatrix} -\tfrac{1}{2} I_{11}^x & -1 \\ -\tfrac{1}{6} I_1^u & \tfrac{1}{2} I_{11}^x \end{pmatrix} \tag{5.5}
$$

where the Replacement Theorem can used to produce the invariantisation of $\varrho_s \varrho^{-1}$ when it is given in terms of $(x, u, x_s, \ldots)$.

It can happen that the infinitesimal vector fields of a transformation group are known, but not the group action, much less a matrix representation of the group. This is the case when investigating symmetry groups of differential equations using Lie's algorithm. Given the Lie bracket multiplication table of the infinitesimal vector fields, a matrix representation of the Lie algebra can always be constructed, a result known as Ado's Theorem; further, the construction has been implemented (de Graaf, 2000). Applications of Theorem 5.2.4 to partial differential equations will be given in Section 5.6, and applications to invariant ordinary differential equations will be given in Chapter 6. The analogue of Theorem 5.2.4 in terms of the infinitesimal vector fields, without reference to a matrix representation, is given in Section 6.7.

The following proposition generalises a result, well known in many different contexts, to the case of non-commuting invariant differential operators. In Differential Geometry, this result is essentially the structural formula for the Maurer–Cartan form (Choquet-Bruhat and DeWitt-Morette, 1982, page 208), while in physical gauge theories, it is the zero curvature equation.

Proposition 5.2.8 *The curvature matrices (5.2) satisfy the syzygy*

$$
\mathcal{D}_j(\mathcal{Q}_i) - \mathcal{D}_i(\mathcal{Q}_j) = ([\mathcal{D}_j, \mathcal{D}_i]\varrho)\varrho^{-1} + [\mathcal{Q}_j, \mathcal{Q}_i]. \tag{5.6}
$$

Exercise 5.2.9 Prove Proposition 5.2.8. Hint: $\varrho\varrho^{-1} = 1$ implies $\mathcal{D}_k(\varrho^{-1}) = -\varrho^{-1}\mathcal{D}_k(\varrho)\varrho^{-1}$.

If $p > 1$, the components of equation (5.6) yield first order syzygies of the symbolic invariants.

Example 5.2.10 We take the group $SO(2) \ltimes \mathbb{R}^2$ with representation

$$\mathcal{R}(a, b, \theta)^{-1} = \begin{pmatrix} \cos\theta & \sin\theta & -b\sin\theta - a\cos\theta \\ -\sin\theta & \cos\theta & -b\cos\theta + a\sin\theta \\ 0 & 0 & 1 \end{pmatrix}$$

acting on $(x, t, u(x, t))$-space as $\tilde{t} = t$ and

$$\begin{pmatrix} \tilde{x} \\ \tilde{u} \end{pmatrix} = \begin{pmatrix} \cos\theta & \sin\theta \\ -\sin\theta & \cos\theta \end{pmatrix} \begin{pmatrix} x - a \\ u - b \end{pmatrix}.$$

Since we use the inverse action we take the inverse of the standard representation. The generators of the matrix representation of the Lie algebra are

$$\mathfrak{a}_a = \begin{pmatrix} 0 & 0 & -1 \\ 0 & 0 & 0 \\ 0 & 0 & 0 \end{pmatrix}, \qquad \mathfrak{a}_b = \begin{pmatrix} 0 & 0 & 0 \\ 0 & 0 & -1 \\ 0 & 0 & 0 \end{pmatrix}$$

and

$$\mathfrak{a}_\theta = \begin{pmatrix} 0 & 1 & 0 \\ -1 & 0 & 0 \\ 0 & 0 & 0 \end{pmatrix}.$$

If we take the normalisation equations $\tilde{x} = 0$, $\tilde{u} = 0$ and $\tilde{u}_x = 0$, the correction matrix is

$$\mathsf{K} = \begin{matrix} & a & b & \theta \\ x & \begin{pmatrix} 1 & 0 & I_{11}^u \\ 0 & I_2^u & I_{12}^u \end{pmatrix} \end{matrix}$$

and hence

$$\mathcal{Q}_x = \begin{pmatrix} 0 & I_{11}^u & -1 \\ -I_{11}^u & 0 & 0 \\ 0 & 0 & 0 \end{pmatrix}, \qquad \mathcal{Q}_t = \begin{pmatrix} 0 & I_{12}^u & 0 \\ -I_{12}^u & 0 & -I_2^u \\ 0 & 0 & 0 \end{pmatrix}.$$

Setting these into equation (5.6), from the $(1, 3)$ component we can deduce that

$$[\mathcal{D}_x, \mathcal{D}_y] = -I_{11}^u I_2^u \mathcal{D}_x,$$

if we had not calculated it already, while the $(2, 3)$ and $(1, 2)$ components yield respectively

$$\mathcal{D}_x I_2^u = I_{12}^u, \qquad \mathcal{D}_t I_{11}^u - \mathcal{D}_x I_{12}^u = I_2^u (I_{11}^u)^2.$$

These last two combine to yield the syzygy between the generators, I_2^u and I_{11}^u, namely

$$\mathcal{D}_t I_{11}^u = \left(\mathcal{D}_x^2 + (I_{11}^u)^2\right) I_2^u. \tag{5.7}$$

Other fully worked examples can be found in Mansfield and van der Kamp (2006). Applications of syzygies will appear in Section 5.6 and in Chapter 7.

5.3 Notes for symbolic computation

In a symbolic computing environment, one often wants to simplify invariant expressions with respect to the invariantised normalisation equations and more generally the syzygies. Typically the normalisation equations are polynomials in their arguments, and then, in order to obtain well-defined reduction processes, one must assume that the normalisation equations are an algebraic Gröbner basis for the algebraic ideal they generate. It should be borne in mind, however, that not all Gröbner bases are suitable to define a frame. The solution space of the equations $\Psi = 0$ needs to be a unique surface of dimension equal to that of the ambient space less the dimension of the group. But this is not enough; normalisation equations such as $(I_1)^2 = 0$ will lead to undetected zero coefficients of leading terms, and even zero denominators. So, the normalisation equations need to form a prime ideal.

For a given term ordering on the set $\{I_K^\alpha \mid |K| \geq 0\}$, a normalisation equation $\psi(J_i, I^\alpha, I_K^\alpha) = 0$ (containing terms other than J_j, $j = 1, \ldots, p$) will have a leading invariant term. This term is called a 'highest normalised invariant'. Denote the set of such highest normalised invariants by $\mathcal{H}NI$. Similarly, we may take an ordering on the terms J_j, $j = 1, \ldots, p$ and obtain from those normalisation equations not containing any of the I_K^α, a set $\mathcal{H}J$ of highest normalised invariants deriving from the independent variables.

Simplification problems noted above are eliminated if we assume that the elements of $\mathcal{H}NI$ and $\mathcal{H}J$ occur linearly in the normalisation equations. In this case standard simplification procedures have the effect of eliminating the highest normalised invariants from all results of all calculations.

Simplifying with respect to differential syzygies in a well-defined way is a much harder problem. To begin with, depending on the group action, the correction terms M_{Kj}^α may not be of polynomial type. If the group action is rational and the normalisation equations are polynomials, the correction terms may contain denominators. Fortunately, the multiplicative set generated by factors of the possible denominators can be determined in advance; they are the factors of the denominators appearing in the matrix **K**.

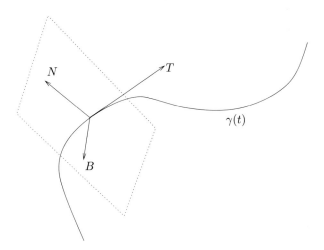

Figure 5.6 An orthonormal 'frame' of vectors defined at every point of a curve in $\mathbb{R}^3$; T is the unit tangent, while N and B lie in the plane normal to T.

The difficulties of forging a theory of differential algebra for the symbolic invariant calculus, that gives well defined simplification processes in a symbolic computation environment, for large classes of Lie group actions, were overcome by Hubert (2005). The difficulty that sets the invariant calculus apart from earlier work in differential algebra (Hubert, 2000) is the existence of non-commuting differential operators. An exposition of Hubert's 2005 paper is beyond the scope of the present book, but it is important to be aware of the difficulties of reduction processes in symbolic computation, and the existence of a solution.

5.4 *The Serret–Frenet frame

This classical example concerns the linear action of the Euclidean group $SE(3)$ of rotations and translations on curves in $\mathbb{R}^3$. Since an element of $SO(3)$ is equivalent to an orthonormal basis of $\mathbb{R}^3$, and since the tangent space at a point of a curve is isomorphic to $\mathbb{R}^3$, one can think of the moving frame, as a map from the jet space on curves to $SO(3)$, literally as a 'frame' of orthonormal vectors at each point of the curve, see Figure 5.6. This conflation of the classical and modern concepts of frames applies particularly to examples arising in Differential Geometry.

We will first derive the Serret–Frenet equations and then show how they may be obtained using Theorem 5.2.4. This example is thus one of several we calculate to demonstrate Theorem 5.2.4. In this particular case, calculating everything needed to apply the theorem looks more complicated than obtaining the Serret–Frenet equations directly, mainly because we are giving complete details in this pedagogic example, and using the opportunity to show how the new and the classical ideas are related.

Remark 5.4.1 The value of Theorem 5.2.4 is not that it can be used to reproduce well-known results. The value lies in that the curvature matrices can be obtained algorithmically in a symbolic computation environment once the action and the normalisation equations are given; no geometric insight is needed. Algorithmic methods allow one to play with the normalisation equations and representations to find those that offer particular computational advantages in the application at hand.

The method shown in this example can be adapted to a wide variety of linear Lie group actions acting on tangent spaces.

Notation We use $\langle \mathbf{v}, \mathbf{w} \rangle$ for the scalar product of vectors $\mathbf{v}$ and $\mathbf{w}$, and $\| \mathbf{v} \| = \sqrt{\langle \mathbf{v}, \mathbf{v} \rangle}$.

Let a curve $s \mapsto \gamma(s) \subset \mathbb{R}^3$ be given. If s is arc length, then we have

$$\langle \gamma'(s), \gamma'(s) \rangle = 1 \tag{5.8}$$

for all s. This follows from the three dimensional Pythagoras Theorem, which implies $\Delta s^2 \approx (\Delta x)^2 + (\Delta y)^2 + (\Delta z)^2$. The standard definitions of T, N and B, the tangent vector, the normal vector and the binormal vector at some fixed point on the curve, are

$$T = \gamma_s, \qquad N = \frac{\gamma_{ss}}{\| \gamma_{ss} \|}, \qquad B = T \times N.$$

From equation (5.8) we have $\| T \| = 1$ and $\langle T, N \rangle = 0$, so that the three vectors T, N and B form an orthonormal set, see Figure 5.6.

Definition 5.4.2 The definition of the *Euclidean curvature* of the curve $\gamma(t) \subset \mathbb{R}^n$ at each point is

$$\kappa = \frac{\| \gamma_t \times \gamma_{tt} \|}{\| \gamma_t \|^3}. \tag{5.9}$$

The *torsion* is

$$\tau = \frac{\langle \gamma_t \times \gamma_{tt}, \gamma_{ttt} \rangle}{\| \gamma_t \times \gamma_{tt} \|^2}. \tag{5.10}$$

For a curve parametrised by arc length s, $T = \gamma_s$ satisfies $\langle T, T \rangle = 1$, so that

$$\kappa = \| T_s \| = \| \gamma_{ss} \|, \tag{5.11}$$

while the *torsion* becomes

$$\tau = \frac{1}{\kappa^2} \langle T \times T_s, T_{ss} \rangle. \tag{5.12}$$

Exercise 5.4.3 Prove that the curvature and torsion are differential invariants of the standard action of $SE(3)$ on curves in $\mathbb{R}^3$.

Theorem 5.4.4 *The Serret–Frenet equations for a curve parametrised by arc length are*

$$\begin{aligned} T_s &= \kappa N \\ N_s &= -\kappa T + \tau B \\ B_s &= -\tau N. \end{aligned} \tag{5.13}$$

Proof The first equation follows from the definition of the curvature κ. To obtain the other two Serret–Frenet equations, we first note that from $\langle T, T \rangle = 1$ we have $\|T_s\|^2 + \langle T_{ss}, T \rangle = 0$. Differentiating $T_s = \kappa N$ yields

$$N_s = \frac{1}{\kappa} T_{ss} - \frac{\kappa_s}{\kappa} N \tag{5.14}$$

so that $\langle N_s, T \rangle = -\kappa$. Next,

$$\langle N_s, T \times N \rangle = \frac{1}{\kappa^2} \langle T_{ss}, T \times T_s \rangle = \tau,$$

while $\langle N, N \rangle = 1$ implies $\langle N_s, N \rangle = 0$. Since any vector is determined by its components in the directions T, N and B, the second Serret–Frenet equation results. The third follows from the first two by noting that $B_s = T_s \times N + T \times N_s$. $\qquad\square$

Exercise 5.4.5 Show using equation (5.14) that $\langle T_{ss}, N \rangle = \kappa_s$.

Exercise 5.4.6 If the curve is not parametrised by arc length, so that $\langle \gamma_t, \gamma_t \rangle = v$ is not necessarily unity, show that

$$\begin{aligned} T_t &= v \kappa N \\ N_t &= -v \kappa T + v \tau B \\ B_t &= -v \tau N \end{aligned}$$

where

$$T = \frac{\gamma_t}{\|\gamma_t\|}, \qquad N = B \times T, \qquad B = \frac{\gamma_t \times \gamma_{tt}}{\|\gamma_t \times \gamma_{tt}\|}.$$

If we put T, N and B to be the columns of a matrix g, we have that

$$g = (T \ N \ B) \in SO(3). \tag{5.15}$$

The matrix in equation (5.15) is known as the classical Serret–Frenet frame. Indeed, since the columns are orthonormal, $g^T g = I_3$, the 3×3 identity matrix, so that $g^{-1} = g^T$. Also,

$$g^{-1} g_s = \begin{pmatrix} 0 & \kappa & 0 \\ -\kappa & 0 & \tau \\ 0 & -\tau & 0 \end{pmatrix} \in \mathfrak{so}(3) \qquad (5.16)$$

by the Serret–Frenet equations. This provides a different proof that κ and τ are rotation invariants.

Our aim in the rest of this section is to show how equation (5.16) is of the form $\rho_s \rho^{-1} = A$ where A is given by Theorem 5.2.4.

The standard left action of $SE(3) = SO(3) \ltimes \mathbb{R}^3$ in $\mathbb{R}^3$ is

$$(R, \mathbf{v}) * \mathbf{x} = R\mathbf{x} + \mathbf{v}$$

and this prolongs to curves $\mathbf{x}(s) \subset \mathbb{R}^3$ as

$$(R, \mathbf{v}) * (\mathbf{x}, \mathbf{x}_s, \mathbf{x}_{ss}, \dots) = (R\mathbf{x} + \mathbf{v}, R\mathbf{x}_s, R\mathbf{x}_{ss}, \dots)$$

since

$$\frac{\mathrm{d}}{\mathrm{d}s}(R\mathbf{x} + \mathbf{v}) = R\mathbf{x}_s$$

and so forth. Thus the induced action on the matrix $g = (T \ N \ B)$ where the vectors T, N and B are columns is

$$(R, \mathbf{v}) * g = (R, \mathbf{v}) * (T \ N \ B) = (RT \ RN \ RB) = Rg$$

and similarly,

$$(R, \mathbf{v}) * g_s = (R, \mathbf{v}) * (T_s \ N_s \ B_s) = (RT_s \ RN_s \ RB_s) = Rg_s.$$

In other words, the action on the vectors associated with the curve is precisely that of left multiplication of the group $SO(3)$ on itself, and the prolonged action is the same as that induced by left multiplication on the tangent space of the group $TSO(3)$. This is precisely the case considered in Example 4.4.12. In that example, we defined a frame $\rho(g) = g^{-1}$ from the normalisation equation $\tilde{g} = e$, and so ρ is 'our Serret–Frenet frame'. Since $(g^{-1})_s = -g^{-1}g_s g^{-1}$, it follows that $\rho_s \rho^{-1} = -g^{-1}g_s$.

We now restrict our attention to the action on the coordinates x_s, y_s and z_s and their derivatives, that are equivalent to that for the frame $\rho(g) = g^{-1}$ above. We will then show that $\rho_s \rho^{-1}$ is indeed the negative of the matrix appearing in the Serret–Frenet frame equations.

In the case at hand, the normalisation equation $\tilde{g} = e$ amounts to $\tilde{T} = (1 \ 0 \ 0)^T$ and $\tilde{N} = (0 \ 1 \ 0)^T$, from which $\tilde{B} = (0 \ 0 \ 1)^T$ follows. In terms of the original curve $\mathbf{x} = (x, y, z)$, we have since $T = (x_s \ y_s \ z_s)^T$ that the three normalisation equations for the three rotation parameters, giving 'our' Serret–Frenet frame, are

$$\widetilde{y}_s = \widetilde{z}_s = \widetilde{z}_{ss} = 0. \tag{5.17}$$

Let θ_{ab} be the parameter of rotation about the origin in the (a, b)-plane, and $\mathbf{v}_{ab}$ the associated infinitesimal vector field. The infinitesimal vector fields of the rotation group are

$$\begin{aligned}
\mathbf{v}_{xy} &= y\partial_x - x\partial_y \\
\mathbf{v}_{xz} &= z\partial_x - x\partial_z \\
\mathbf{v}_{yz} &= z\partial_y - y\partial_z
\end{aligned} \tag{5.18}$$

so that the table of infinitesimals is

$$
\begin{array}{c}
\begin{array}{ccccccc}
s & x & y & z & y_s & z_s & z_{ss}
\end{array} \\
\begin{array}{c}
\theta_{xy} \\
\theta_{xz} \\
\theta_{yz}
\end{array}
\left(
\begin{array}{ccccccc}
0 & y & -x & 0 & -x_s & 0 & 0 \\
0 & z & 0 & -x & 0 & -x_s & -x_{ss} \\
0 & 0 & z & -y & z_s & -y_s & -y_{ss}
\end{array}
\right).
\end{array}
$$

The matrix $\mathbf{K}$ is thus

$$
\mathbf{K} = s
\begin{array}{c}
\begin{array}{ccc}
\theta_{xy} & \theta_{xz} & \theta_{yz}
\end{array} \\
\left(
\begin{array}{ccc}
I_{11}^{y} & 0 & I_{111}^{z} \\
I_{1}^{x} & & I_{11}^{y}
\end{array}
\right).
\end{array}
$$

The Lie bracket multiplication table for the infinitesimal vector fields is

$$
\begin{array}{c|ccc}
[\,,\,] & \mathbf{v}_{xy} & \mathbf{v}_{xz} & \mathbf{v}_{yz} \\
\hline
\mathbf{v}_{xy} & 0 & \mathbf{v}_{yz} & -\mathbf{v}_{xz} \\
\mathbf{v}_{xz} & -\mathbf{v}_{yz} & 0 & \mathbf{v}_{xy} \\
\mathbf{v}_{yz} & \mathbf{v}_{xz} & -\mathbf{v}_{xy} & 0
\end{array} \tag{5.19}
$$

If we take a faithful representation of this Lie algebra to be

$$
A_{xy} =
\begin{pmatrix}
0 & -1 & 0 \\
1 & 0 & 0 \\
0 & 0 & 0
\end{pmatrix}, \quad
A_{xz} =
\begin{pmatrix}
0 & 0 & 1 \\
0 & 0 & 0 \\
-1 & 0 & 0
\end{pmatrix}, \quad
A_{yz} =
\begin{pmatrix}
0 & 0 & 0 \\
0 & 0 & -1 \\
0 & 1 & 0
\end{pmatrix}
$$

we note its Lie bracket table

$$
\begin{array}{c|ccc}
[\,,\,] & A_{xy} & A_{xz} & A_{yz} \\
\hline
A_{xy} & 0 & -A_{yz} & A_{xz} \\
A_{xz} & A_{yz} & 0 & -A_{xy} \\
A_{yz} & -A_{xz} & A_{xy} & 0
\end{array} \tag{5.20}
$$

is the negative of the Lie bracket table for the vector fields, (5.19). Hence by Theorem 5.2.4 and noting Remark 5.2.5, we have

$$\rho_s \rho^{-1} = \mathbf{K}_{s\theta(xy)} A_{xy} + \mathbf{K}_{s\theta(xz)} A_{xz} + \mathbf{K}_{s\theta(yz)} A_{yz}$$

$$= \begin{pmatrix} 0 & -\dfrac{I_{11}^y}{I_1^x} & 0 \\[2ex] \dfrac{I_{11}^y}{I_1^x} & 0 & -\dfrac{I_{111}^z}{I_{11}^y} \\[2ex] 0 & \dfrac{I_{111}^z}{I_{11}^y} & 0 \end{pmatrix}.$$

If s is arc length, then $\rho T = (1\,0\,0)^T = (I_1^x\ I_1^y\ I_1^z)^T$ by the Fels–Olver–Thomas Replacement Theorem, so we have $I_1^x = 1$. Next, since

$$\rho N = \rho T_s / \kappa = \rho (x_{ss}\ y_{ss}\ z_{ss})^T = \frac{1}{\kappa}(I_{11}^x\ I_{11}^y\ I_{11}^z)^T = (0\,1\,0)^T$$

we have $\kappa = I_{11}^y$ (and also $I_{11}^z = 0$, which follows as well from the invariant differentiation formulae applied to $I_1^x = 1$). Finally,

$$\begin{vmatrix} x_s & y_s & z_s \\ x_{ss} & y_{ss} & z_{ss} \\ x_{sss} & y_{sss} & z_{sss} \end{vmatrix} = \begin{vmatrix} I_1^x & 0 & 0 \\ I_{11}^x & I_{11}^y & 0 \\ I_{111}^x & I_{111}^y & I_{111}^z \end{vmatrix} = I_1^x I_{11}^y I_{111}^z.$$

Hence

$$\tau = \frac{I_1^x I_{11}^y I_{111}^z}{(I_{11}^y)^2} = \frac{I_{111}^z}{I_{11}^y}.$$

Thus

$$\rho_s \rho^{-1} = \begin{pmatrix} 0 & -\kappa & 0 \\ \kappa & 0 & -\tau \\ 0 & \tau & 0 \end{pmatrix}$$

as promised.

Exercise 5.4.7 Redo the calculation above but with a scaling action $\tilde{s} = \lambda s$ on the curve parameter. Use the normalisation equation $I_1^x = 1$. Can this be used to mimic the effect of s being the arc length, or are additional equations still required? Hint: does the invariant $I_{11}^x = 0$?

We will use the result of the next exercise in Section 5.6.

Exercise 5.4.8 Redo the above calculation but this time applying Example 4.4.12 to the full rotation and translation group, $SE(3)$. Use the normalisation equations $\tilde{x} = \tilde{y} = \tilde{z} = 0$, that is, the generic point on the curve is translated to the origin, as well as those used previously, given in equation (5.17). Suppose that not only is the curve parametrised by (arc length) s, but also that it evolves in time t. Taking $\tilde{s} = s, \tilde{t} = t$, show the matrix $\mathbf{K}$ is now

$$
\mathbf{K} = \begin{array}{c} \\ s \\ t \end{array}
\begin{array}{cc}
\begin{array}{cccccc}
k_x & k_y & k_z & \theta_{xy} & \theta_{xz} & \theta_{yz}
\end{array} \\
\left(
\begin{array}{cccccc}
-I_1^x & 0 & 0 & \dfrac{I_{11}^y}{I_1^x} & 0 & \dfrac{I_{111}^z}{I_{11}^y} \\[3mm]
-I_2^x & -I_2^y & -I_2^z & \dfrac{I_{12}^y}{I_1^x} & \dfrac{I_{12}^z}{I_1^x} & \dfrac{I_{112}^z}{I_{11}^y} - \dfrac{I_{12}^z I_{11}^x}{I_1^x I_{11}^y}
\end{array}
\right)
\end{array}
$$

where the k_i is the group parameter for translation in the ith direction. Show that if s is arc length, so that $I_1^x = 1$, then $I_{11}^x = I_{12}^x = 0$. Using the standard representation of $SE(3)$ in $GL(4)$, find $\rho_t \rho^{-1}$ and calculate the components of

$$
\frac{\partial}{\partial t} \rho_s \rho^{-1} - \frac{\partial}{\partial s} \rho_t \rho^{-1} = [\rho_t \rho^{-1}, \rho_s \rho^{-1}].
$$

Since both t and s are invariant, we have $\mathcal{D}_t = \partial/\partial t$ and $\mathcal{D}_s = \partial/\partial s$. Show that the resulting syzygies can also be obtained via the invariant differentiation formulae. Draw the diagrams of invariants similar to that in Figure 5.3 for this action and frame; there will be one for each of x, y and z, but the invariants I_{1K}^x may all be determined in terms of the others by virtue of $I_1^x = 1$ when s is arc length. Show that the generating syzygies can be written in the form

$$
\frac{\partial}{\partial t} \begin{pmatrix} I_1^x \\ \kappa \\ \tau \end{pmatrix} = \mathcal{H} \begin{pmatrix} I_2^x \\ I_2^y \\ I_2^z \end{pmatrix} \tag{5.21}
$$

where $\mathcal{H}$ is a matrix of differential operators with respect to s only, whose coefficients depend only on κ, τ and I_1^x, and their derivatives with respect to s only.

Exercise 5.4.9 There is a different frame for the Euclidean action on curves in $\mathbb{R}^3$ called the *normal frame*. The vector T is the same as for the Serret–Frenet frame, but there is a different choice of N and B such that setting $g = (T \ N \ B) \in SO(3)$, one has

$$
g^{-1} g_s = \begin{pmatrix} 0 & -k_1 & -k_2 \\ k_1 & 0 & 0 \\ k_2 & 0 & 0 \end{pmatrix}. \tag{5.22}
$$

Let $\mathbf{x}$ be a point on the curve, and consider the rotation in the plane normal to T about the point $\mathbf{x}$, that is, about the origin in $T_\mathbf{x}\mathbb{R}^3$, that takes the Serret–Frenet frame ρ_{SF} to the normal frame ρ_N. Show $\rho_N \rho_{SF}^{-1}$ is a rotation in the (y, z)

plane about the x-axis. Show the associated rotation parameter $\theta(s)$ satisfies a differential equation, $\theta_s = \tau$, and that $k_1 = \kappa \cos\theta$, $k_2 = \kappa \sin\theta$. Can the normal frame be obtained via normalisation equations which are functions of x_s, y_s, z_s, and their derivatives?

5.5 *Curvature matrices for linear actions

In this section we examine linear actions of matrix Lie groups acting on curves in $\mathbb{R}^n$. Our study of this case was begun in Section 4.5.1; we repeat the basic facts for convenience.

We denote the curve as a column vector $\mathbf{u} = \left(u^1(s), u^2(s), \ldots, u^n(s)\right)^T$ and take the affine action of $G \ltimes \mathbb{R}^n$ as

$$(g, \mathbf{v}) \cdot \mathbf{u} = g\mathbf{u} + \mathbf{v}.$$

If we denote the nth derivative of $\mathbf{u}$ with respect to s as

$$\frac{d^n \mathbf{u}}{ds^n} = \mathbf{u}_n$$

then

$$(g, \mathbf{v}) \cdot \mathbf{u}_n = g\mathbf{u}_n, \qquad n > 0.$$

We recall from Section 4.5.1 the formula for the symbolic differentiation of the invariantised derivatives $\iota(\mathbf{u}_k)$. If ρ is a frame so that $\rho\mathbf{u}_k = \iota(\mathbf{u}_k)$, then equation (4.18) gives

$$\frac{d}{ds}\iota(\mathbf{u}_k) = \iota(\mathbf{u}_{k+1}) + \rho_s\rho^{-1}\iota(\mathbf{u}_k). \tag{5.23}$$

Typically, the translation part of the affine group action is normalised by $\widetilde{\mathbf{u}} = 0$, and one then is restricted to considering the linear action of G on $\mathbf{u}_i$, $i \geq 1$. The frame ρ that one considers is then an equivariant map on the space of first and higher order derivatives of the curve, to G.

The main idea of this section is to show how one might go about choosing the remaining normalisation equations so that $\rho_s\rho^{-1}$ has the property that its non-constant components are functionally distinct invariants. The main example we investigate is $G = Sp(2)$. The result for $Sp(n)$, $n \geq 2$, amongst other examples, is detailed in Marí Beffa (2008a).

Exercise 5.5.1 Let the Lie group G be of the form

$$G = \{A \in GL(n) \mid A^T S A = S\} \tag{5.24}$$

for some specified $n \times n$ matrix S. Show that the scalar functions

$$\mathbf{u}_k^T S \mathbf{u}_\ell$$

where $k, \ell > 0$, are differential invariants. Further, if we define the $n \times n$ matrix U as

$$U = \begin{pmatrix} \mathbf{u}_1 \ \mathbf{u}_2 \ldots \mathbf{u}_n \end{pmatrix}, \tag{5.25}$$

then

$$U^T S U$$

has all its components invariant. If ρ is a frame for the action, so that $\iota(U) = \rho U$, show that

$$\iota(U)^T S \iota(U) = U^T \rho^T S \rho U = U^T S U. \tag{5.26}$$

If $S^T = -S$ then show that

$$\frac{\mathrm{d}}{\mathrm{d}s} \left(\mathbf{u}_{k+1}^T S \mathbf{u}_k \right) = \mathbf{u}_{k+2}^T S \mathbf{u}_k.$$

Consider $G = Sp(2) \subset GL(4, \mathbb{R})$ where $Sp(2)$ is the symplectic group defined in Exercise 3.2.21. The symplectic group is of the form in equation (5.24), where $S = J$ and where J is the 4×4 matrix

$$J = \begin{pmatrix} 0 & I_2 \\ -I_2 & 0 \end{pmatrix}$$

and I_2 is the 2×2 identity matrix. We have $J^T = -J$ and thus all the results of Exercise 5.5.1 hold.

If the matrix U is given as in equation (5.25), with $n = 4$ since $Sp(2) \subset GL(4)$, then the normalisation equations giving a frame for the action on curves impose conditions on the entries of $\iota(U)$. Since the dimension of $Sp(2)$ is ten, we need ten conditions.

Exercise 5.5.2 Setting

$$U = \begin{pmatrix} \mathbf{u}_1 & \mathbf{u}_2 & \mathbf{u}_3 & \mathbf{u}_4 \end{pmatrix}$$

and

$$\kappa_1 = \mathbf{u}_2^T J \mathbf{u}_1, \qquad \kappa_2 = \mathbf{u}_3^T J \mathbf{u}_2, \qquad \kappa_3 = \mathbf{u}_4^T J \mathbf{u}_3,$$

show

$$U^T J U = \begin{pmatrix} 0 & -\kappa_1 & -\kappa_1' & -(\kappa_1'' - \kappa_2) \\ \kappa_1 & 0 & -\kappa_2 & -\kappa_2' \\ \kappa_1' & \kappa_2 & 0 & -\kappa_3 \\ \kappa_1'' - \kappa_2 & \kappa_2' & \kappa_3 & 0 \end{pmatrix} = \kappa,$$

where we have denoted d/ds by a prime, $'$, and where this defines the matrix κ.

To begin the normalisation process, we set the first four normalisation equations to be $\widetilde{\mathbf{u}}_1 = (1\ 0\ 0\ 0)^T$. This means that

$$\iota(\mathbf{u}_1) = \begin{pmatrix} 1 \\ 0 \\ 0 \\ 0 \end{pmatrix}$$

and since $\iota(U) = \rho U$ satisfies

$$\iota(U)^T J \iota(U) = U^T J U = \kappa, \tag{5.27}$$

we obtain conditions on the next column, $\iota(\mathbf{u}_2)$ of $\iota(U)$. Doing the calculation reveals that we may choose all but one of the components of $\widetilde{\mathbf{u}}_2$ to be zero, but that we must have that $\iota(u_2^3) = -\kappa_1$ for equation (5.27) to hold. Hence we take the next three normalisation equations to be

$$\widetilde{u}_2^1 = \widetilde{u}_2^2 = \widetilde{u}_2^4 = 0$$

and obtain

$$\iota(\mathbf{u}_2) = \begin{pmatrix} 0 \\ 0 \\ -\kappa_1 \\ 0 \end{pmatrix}.$$

Setting this into equation (5.27) reveals conditions on the third and fourth columns, $\iota(\mathbf{u}_3)$ and $\iota(\mathbf{u}_4)$ of $\iota(U)$. Specifically, we must have

$$\left(\iota(\mathbf{u}_3)\ \iota(\mathbf{u}_4) \right) = \begin{pmatrix} -\dfrac{\kappa_2}{\kappa_1} & -\dfrac{\kappa_2'}{\kappa_1} \\ \iota(u_3^2) & \iota(u_4^2) \\ -\kappa_1' & -\kappa_1'' + \kappa_2 \\ \iota(u_3^4) & \iota(u_4^4) \end{pmatrix},$$

where

$$\iota(u_3^2)\iota(u_4^4) - \iota(u_4^2)\iota(u_3^4) = \frac{\kappa_1'\kappa_2'}{\kappa_1} + \frac{\kappa_2}{\kappa_1}\left(\kappa_2 - \kappa_1''\right) - \kappa_3. \tag{5.28}$$

We have three undetermined components of $\iota(U)$ and three remaining normalisation equations to be chosen. So far, we have not looked at the form that $\rho_s \rho^{-1}$ takes, and to do this we note that we have both equation (5.23) to satisfy, and that $\rho_s \rho^{-1} \in \mathfrak{sp}(2)$. The form that elements of $\mathfrak{sp}(2)$ take was calculated in Exercise 3.2.21; they depend on ten parameters and are of the form

$$\rho_s \rho^{-1} = \begin{pmatrix} a_1 & a_2 & b_1 & b_2 \\ a_3 & a_4 & b_2 & b_3 \\ c_1 & c_2 & -a_1 & -a_3 \\ c_2 & c_3 & -a_2 & -a_4 \end{pmatrix}.$$

Setting this together with the elements of $\iota(U)$ obtained so far into equation (5.23), we obtain

$$a_1 = a_3 = c_2 = 0, \qquad a_2 = -\frac{\iota(u_3^4)}{\kappa_1}, \qquad b_2 = \frac{\iota(u_3^2)}{\kappa_1}, \qquad c_1 = \kappa_1. \qquad (5.29)$$

We now take note that we want the components of $\rho_s \rho^{-1}$ to be either constant or functionally independent invariants. Taking into account equations (5.28) and (5.29), we may choose two more normalisation equations from the components of $\widetilde{\mathbf{u}}_3$,

$$\widetilde{u}_3^4 = 0, \qquad \widetilde{u}_3^2 = \kappa_1.$$

Inserting all the above information and choices, into equations (5.23), (5.27) and (5.28) we have that $\iota(u_4^4)$ is determined, indeed we have

$$\iota(u_4^4) = -\frac{\kappa_2 \kappa_1''}{\kappa_1^2} + \frac{\kappa_2^2}{\kappa_1^2} + \frac{\kappa_2' \kappa_1'}{\kappa_1^2} - \frac{\kappa_3}{\kappa_1}$$

and that

$$c_3 = -\frac{\iota(u_4^4)}{\kappa_1}, \qquad a_4 = -\frac{\iota(u_4^2)}{\kappa_1} + 2\frac{\kappa_1'}{\kappa_1}.$$

With all this, we have that so far

$$\rho_s \rho^{-1} = \begin{pmatrix} 0 & 0 & -\dfrac{\kappa_2}{\kappa_1^2} & 1 \\ 0 & \dfrac{1}{\kappa_1}\left(2\kappa_1' - \iota(u_4^2)\right) & 1 & b_3 \\ \kappa_1 & 0 & 0 & 0 \\ 0 & \dfrac{1}{\kappa_1^3}\left(\kappa_2 \kappa_1'' - \kappa_2^2 - \kappa_2' \kappa_1' + \kappa_3 \kappa_1\right) & 0 & -\dfrac{1}{\kappa_1}\left(2\kappa_1' - \iota(u_4^2)\right) \end{pmatrix}$$

where b_3 may be determined from a fifth order invariant using equation (5.23). In order to have that the non-constant components of $\rho_s \rho^{-1}$ are functionally distinct, the tenth and final normalisation equation is taken to be

$$\widetilde{u}_4^2 = 2\kappa_1'.$$

Finally, one has that

$$b_3 = -\frac{\kappa_1^2 \left(\iota(u_5^2) + \kappa_2 - 3\kappa_1'' \right)}{\kappa_2^2 + \kappa_1' \kappa_2' - \kappa_2 \kappa_1'' - \kappa_1 \kappa_3}$$

while the other fifth order invariants are determined to be

$$\iota(u_5^1) = \frac{\kappa_3 - \kappa_2''}{\kappa_1}$$
$$\iota(u_5^3) = 2\kappa_2' - \kappa_1'''$$
$$\iota(u_5^4) = -\frac{1}{\kappa_1^2} \left(\kappa_3' \kappa_1 + \kappa_2 \kappa_1''' - 2\kappa_2 \kappa_2' - \kappa_2'' \kappa_1' + \kappa_3 \kappa_1' \right).$$

It follows from equation (5.23) that the sixth and higher order invariants may be determined from κ_1, κ_2, κ_3 and $\iota(u_5^2)$ and their derivatives.

Exercise 5.5.3 Verify the calculations detailed above. Find an expression for $\iota(u_5^2)$ in terms of κ_1, κ_2, κ_3 and $\kappa_4 = \mathbf{u}_5^T J \mathbf{u}_4$. Hint: use the Replacement Theorem. Conclude that κ_1, κ_2, κ_3 and κ_4 are a finite set of generators and write $\rho_x \rho^{-1}$ in terms of them.

Marí Beffa (2008a) goes on to notice that setting k_j to be the $(j+1)$st order differential invariant appearing in $\rho_s \rho^{-1}$, then

$$\rho_s \rho^{-1} = \begin{pmatrix} 0 & 0 & k_2 & 1 \\ 0 & 0 & 1 & k_4 \\ k_1 & 0 & 0 & 0 \\ 0 & k_3 & 0 & 0 \end{pmatrix}$$

and then shows that in general, for $G = Sp(2n)$ for a suitable choice of normalisation equations, one may obtain the beautiful result,

$$\rho_s \rho^{-1} = \left(\begin{array}{cccc|cccc} & & & & k_2 & 1 & 0 & & 0 \\ & & & & 1 & k_4 & 1 & & 0 \\ & & 0 & & & & \ddots & & \\ & & & & & & 1 & k_{2n-2} & 1 \\ & & & & 0 & & 0 & 1 & k_{2n} \\ \hline k_1 & 0 & & 0 & & & & & \\ 0 & k_3 & & & & & & & \\ & & \ddots & & & & 0 & & \\ 0 & & & k_{2n-1} & & & & & \end{array} \right).$$

5.6 *Curvature flows

As ever, the favourite example comes from an $SL(2)$ action. For the projective action

$$\widetilde{x} = x, \qquad \widetilde{t} = t, \qquad \widetilde{u} = \frac{au + b}{cu + d}, \qquad ad - bc = 1,$$

the generating invariants are

$$I_{111}^u = \{u; x\} = \frac{u_{xxx}}{u_x} - \frac{3}{2}\left(\frac{u_{xx}}{u_x}\right)^2, \qquad I_2^u = \frac{u_t}{u_x}.$$

Setting $V = I_{111}^u$, $W = I_2^u$, the syzygy is

$$\frac{\partial}{\partial t} V = \left(\frac{\partial^3}{\partial x^3} + 2V\frac{\partial}{\partial x} + V_x\right)W. \tag{5.30}$$

The operator on the right hand side of equation (5.30) is well known as one of the Hamiltonian operators associated with the Korteweg–de Vries equation; if the function $u(x, t)$ satisfies the equation $V = W$, that is, $u_t = u_x\{u; x\}$, then $v = V/2$ satisfies the Korteweg–de Vries equation,

$$v_t = v_{xxx} + 6vv_x.$$

This example is one of many that go under the general name 'curvature flows'. The picture is that of curves $u(x)$ in the plane evolving in such a way that if two curves are related by a group action at time $t = 0$, then so are the curves after time t (see Figure 5.7),

$$(g \cdot u_1 = u_2)\Big|_{t=0} \implies g \cdot u_1 = u_2 \text{ for all } t.$$

The first key idea involves the signature curve of a curve under a moving frame.

Definition 5.6.1 Given a Lie group G, an action of G on a manifold M, and a right moving frame $\rho : U \to G$ with domain $U \subset M$, the *signature curve* of a curve $\gamma(s) \subset U$ is the image of γ under ρ.

All curves equivalent to γ under the group action have the same signature. One can think of the signature as being the projection of the curve onto the cross section $\mathcal{K}$ used to define the frame ρ. Since this cross section has coordinates which are the invariants of the action, another way to think of the signature is as the invariantised curve in invariant space.

Exercise 5.6.2 Consider the group $SE(2)$ of translations and rotations in the (x, u)-plane. A smooth curve in the plane $s \mapsto (x(s), u(s))$ yields a curve

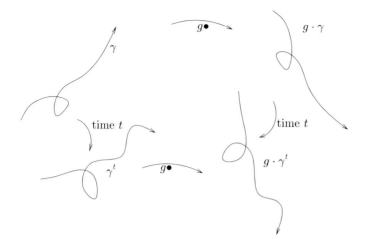

Figure 5.7 An invariant evolution of curves commutes with the group action.

$s \mapsto (x(s), u(s), x_s, u_s, x_{ss}, u_{ss}, \dots)$ in the prolonged space, and hence a curve in the space of Euclidean invariants, $s \mapsto (\kappa, \kappa_s, \dots)$, where

$$\kappa = \frac{u_{ss}x_s - x_{ss}u_s}{(u_s^2 + x_s^2)^{3/2}}.$$

Plot $s \mapsto (\kappa, \kappa_s)$ where κ is Euclidean curvature for parabolas $s \mapsto (s, \alpha s^2 + \beta s + \gamma)$ for various α, β and γ. What do you notice? What are the signature curves for circles? ellipses?

Exercise 5.6.3 For the standard linear action of $SE(3)$, of rotations and translations in (x, y, z)-space, plot the signature curves $s \mapsto (\kappa, \tau)$ for a selection of curves in 3-space, including spirals, under the Euclidean group $SE(3)$; see Definition 5.4.2 for definitions of the curvature κ and the torsion τ.

To ease the exposition, we describe the calculation of the curvature flow equation in the simplest case, curves on the plane. Suppose that a curve $\gamma(s) = (x(s), u(s))$ is evolving in time, and that we have a group action on (x, u) which we prolong to some suitably high order. Then so will the signature curve evolve. If the evolution commutes with the group action, then we can study the evolution of one in terms of the other. To incorporate the time dependence on γ, we could write $\gamma^t(s) = (x^t(s), u^t(s))$, or equivalently $(s, t) \mapsto (x(s, t), u(s, t))$, which is simpler. The evolution of curves is then a system of partial differential equations for x and u. We set $\tilde{t} = t$ and treating x and u as depending on both s and t, can prolong the action to obtain an action on derivatives of x and u with respect to time.

Since there are two independent variables, the error matrix **K** will have two rows from which we obtain the entries to the matrices

$$Q_s = \varrho_s \varrho^{-1}, \qquad Q_t = \varrho_t \varrho^{-1}.$$

Calculating

$$\frac{\partial}{\partial t} Q_s = \frac{\partial}{\partial s} Q_t + [Q_t, Q_s]$$

as in Proposition 5.2.8 leads to syzygies between the 'spatial' invariants obtained by invariantising pure s-derivatives of the dependent variables, and the 'time' invariants obtained by invariantising their pure t-derivatives.

The particular feature of the normalisation equations for moving frames applied to flows on spaces of curves, is that they involve only the dependent variables and their s-derivatives. In the case of curves in the plane parametrised as $(x(s), u(s))$, typically there will be one generating invariant for each of x and u, I_J^x and I_K^u, say. One then finds that the syzygies can be written in the form

$$\frac{\partial}{\partial t} \begin{pmatrix} I_J^x \\ I_K^u \end{pmatrix} = \mathcal{H} \begin{pmatrix} I_2^x \\ I_2^u \end{pmatrix} \tag{5.31}$$

where $\mathcal{H}$ is a matrix of operators depending only on powers of $\partial/\partial s$, the generating s-derivative invariants, I_J^x and I_K^u, and their s-derivatives.

Let the spatial derivative invariants be generated by

$$I_J^x = \eta, \qquad I_K^u = \kappa.$$

A curve evolution which commutes with the group action is standardly of the form, after invariantisation,

$$\begin{aligned} I_2^x &= \mathcal{F}(\eta, \eta_s, \ldots, \kappa, \kappa_s, \ldots) \\ I_2^u &= \mathcal{G}(\eta, \eta_s, \ldots, \kappa, \kappa_s, \ldots) \end{aligned} \tag{5.32}$$

and inserting this into equation (5.31) yields a system of partial differential equations for the generating spatial invariants. If there are more spatial derivative generators than one for each dependent variable, the syzygies between them need to be included in the system. *This system is denoted as the 'curvature flow equations' induced by the flow on curves given by equation (5.32).*

Thus the Korteweg–de Vries equation is the curvature flow corresponding to the flow on curves given by $u_t = u_x\{u; x\}$; this latter equation is called, perhaps unsurprisingly, the 'Schwarzian KdV' equation. Note that reparametrising $(x, u(x))$ as $(x(s), u(s))$ is unnecessary when x is invariant under the group action.

The second most famous example involves the Euclidean action of curves in $\mathbb{R}^3$. Here the flow on curves is the vortex filament equation, and the curvature flow is solved in terms of the non-linear Schrödinger equation. Details are given in Mansfield and van der Kamp (2006).

A third beautiful example[†] is given by examining the standard action of $SL(2) \ltimes \mathbb{R}^2$ on curves in $\mathbb{R}^2$. We now look at the details of this example. The action is $\tilde{s} = s, \tilde{t} = t$ and

$$\begin{pmatrix} \tilde{u} \\ \tilde{v} \end{pmatrix} = \begin{pmatrix} a & b \\ c & d \end{pmatrix} \begin{pmatrix} u \\ v \end{pmatrix} + \begin{pmatrix} k_1 \\ k_2 \end{pmatrix}, \qquad ad - bc = 1.$$

The normalisation equations we take are

$$\tilde{u} = \tilde{v} = \tilde{v}_s = \tilde{u}_{ss} = 0, \qquad \tilde{u}_s = 1.$$

The matrix of infinitesimals is

$$\begin{array}{c} \\ k_1 \\ k_2 \\ a \\ b \\ c \end{array} \begin{array}{cc} u & v \\ \begin{pmatrix} 1 & 0 \\ 0 & 1 \\ u & -v \\ v & 0 \\ 0 & u \end{pmatrix} \end{array}$$

and the error matrix is

$$-\mathbf{K} = \begin{array}{c} \\ s \\ t \end{array} \begin{array}{ccccc} k_1 & k_2 & a & b & c \\ \begin{pmatrix} 1 & 0 & 0 & \dfrac{I^u_{111}}{I^v_{11}} & I^v_{11} \\ I^u_2 & I^v_2 & I^u_{12} & \dfrac{I^u_{112}}{I^v_{11}} & I^v_{12} \end{pmatrix} \end{array}.$$

We consider the case $I^v_{11} = 1$. This is equivalent to considering curves $(v, u(v))$ under the projective action, and taking the curve parameter v to be the 'equi-affine arc length'. Setting

$$1 = I^v_{11}, \qquad \kappa = I^u_{111},$$

the syzygy between I^v_{11} and I^v_2 becomes

$$\frac{\partial^2}{\partial s^2} I^v_2 + 2\kappa I^v_2 + 3\frac{\partial}{\partial s} I^u_2 = 0,$$

which is solved by

$$\begin{aligned} I^v_2 &= \kappa_s \\ I^u_2 &= -\frac{1}{3}(\kappa_{ss} + \kappa^2). \end{aligned} \qquad (5.33)$$

[†] This example was worked out with Gloria Marí Beffa.

This system, together with $I^v_{11} = 1$, is our flow on curves; to obtain it explicitly in terms of u and v, one needs to solve for the frame and evaluate the invariants as functions of u and v.

Putting equations (5.33) into the syzygy between I^u_{111} and I^u_2 yields the curvature flow equation,

$$0 = \kappa_t + \frac{1}{3}\kappa_{5s} + \frac{5}{3}\kappa_s\kappa^2 - \frac{5}{3}\kappa\kappa_{sss} - \frac{5}{3}\kappa_s\kappa_{ss}. \tag{5.34}$$

A simple rescaling of variables, $s \mapsto x$, $t \mapsto 3t$, $\kappa \mapsto -2y$ shows that this is in fact the famous Sawada–Kotera equation (Ablowitz and Clarkson, 1991, page 52),

$$0 = y_t + y_{5x} + 10y_x y_{xx} + 10y y_{xxx} + 20y^2 y_x.$$

Thus far all three curvature flow equations described are well known integrable equations possessing Lax pairs, an infinite number of conservation laws, soliton solutions, and so forth. In fact, if the flow on curves is integrable, then so is the curvature flow equation, and vice versa, in the sense that if one has an infinite number of conserved quantities then so does the other. This is proved in Mansfield and van der Kamp (2006) by showing that the conservation laws factor through the syzygies. The formulation of the syzygies in terms of a matrix $\mathcal{H}$, as in equation (5.31), is pivotal to the discussion.

Much more can be said about moving frames and partial differential equations. We refer the reader to the recent papers of Marí Beffa (2004, 2007, 2008a, 2008b).

6

Invariant ordinary differential equations

We have seen that a group action on a space M induces an action on curves in M and a prolonged action on their tangents and higher order derivatives. Consider now the set of solution curves to an ordinary differential equation, or ODE for short, $\Delta(x, u, u_x, \ldots) = 0$. If G is a symmetry group of Δ, then there is an induced action on the set of solution curves, that is, the action maps solution curves to solution curves. We say two solution curves are equivalent if there is a group element mapping one curve to the other. This is shown in Figure 6.1, which also shows how the frame projects an equivalence class of curves to one curve on the cross section $\mathcal{K}$, drawn here as a surface in M.

The first main result of this chapter is that the moving frame reduces the equation $\Delta(x, u, u_x, \ldots) = 0$ to the system,

$$\Delta(J, I^u, I_1^u, \ldots) = 0$$
$$\mathcal{D}\varrho = \mathcal{Q}\varrho \tag{6.1}$$

where recall $J = \iota(x)$, $I_K^u = \iota(u_K)$, where the first equation is the invariantisation of $\Delta = 0$ obtained using Theorem 4.4.9,

$$\mathcal{D} = \left(\frac{\mathrm{d}\widetilde{x}}{\mathrm{d}x}\right)^{-1} \frac{\mathrm{d}}{\mathrm{d}x}\bigg|_{g=\rho}$$

is the invariantised operator, $\mathcal{Q}$ is the curvature matrix defined in equation (5.2) and ϱ is a faithful matrix representation of the frame ρ. Recall that $\mathcal{Q}$ can be obtained in terms of the symbolic invariants knowing only the infinitesimal action and the normalisation equations, using Theorem 5.2.4.

The first equation solves for how the components of $\mathcal{Q}$ depend on the independent variable x, so the second equation is an equation for ϱ. The system (6.1) is thus *triangular*: the first equation is independent from the second, and the solution of the first is needed to solve the second.

185

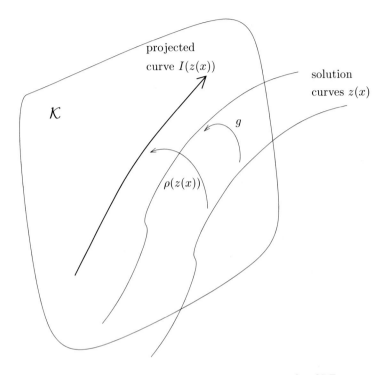

Figure 6.1   The moving frame on the solution curves of an ODE.

Once we have solved for the frame ρ, the function u is obtained from $u = \rho^{-1} \cdot I^u$. To make the calculations tractable, often we need to reparametrise the variables so that the independent variable is an invariant. This ensures that no tricky differential operators are introduced in the equation for the frame. Moreover, we are free to choose the parametrisation to simplify the reduced system as much as possible.

Classical reduction methods are at their most powerful when the symmetry group is solvable, and in this case, the classical reduction process leads to a solution of the ODE by quadratures. Moving frames can be used to advantage in this case since the equation for ϱ, the matrix representation of ρ, can be chosen to be triangular.

But moving frames can be used to handle cases where classical methods give only partial information. To show the power of the ideas while keeping the calculations tractable, we study equations invariant under $SL(2)$. This group is semi-simple, the opposite of solvable, and in these cases we see the advantage of a moving frame approach. Note that Stephani (1989), Chapter 7, gives a method to solve $SL(2)$-invariant ODEs of order 2 or less. The method here is for $SL(2)$ invariant ODEs of order greater than 2. We give a moving frames proof of Schwarz' Theorem and then solve the Chazy equation using moving frames. This last equation was solved by Clarkson and Olver (1996) in a *tour de force*, and the methods there do not generalise to other symmetry groups.

In the final section of this chapter, we discuss how the calculations may be performed using only the vector field presentation of the Lie algebra of the symmetry group, that is, without knowing the group action or a faithful matrix representation. This is important since it is these vector fields that are normally given by symmetry software packages. Integrating the vector fields to obtain the group action, and then finding a matrix representation, can be difficult computations.

6.1 The symmetry group of an ordinary differential equation

Definition 6.1.1 A group G is said to be a *symmetry group* of an ordinary differential system $\Delta = 0$, if the induced action on sufficiently smooth curves maps the set of solution curves of $\Delta = 0$ to itself.

The group action also induces a map on the equations themselves,

$$g \cdot \Delta_i(x, u, u_x, \dots) = \Delta_i(g \cdot x, g \cdot u, \dots).$$

If each Δ_i is a differential invariant of the prolonged action of G, then G is a symmetry group of $\Delta = 0$. More generally, G is a symmetry group of $\Delta = 0$ if there is an invertible matrix (or non-zero function in the case of a single equation) $\mu(g, x, u, u_x, \dots)$ such that

$$\Delta(g \cdot x, g \cdot u, \dots) = \mu(g, x, u, u_x, \dots)\Delta(x, u, u_x, \dots) \qquad (6.2)$$

in which case we say that Δ is a *relative invariant system*.

Example 6.1.2 The solution space of $u_{xx} = 0$ is the set of non-vertical straight lines in the plane. The well-known symmetry group of this equation is the eight dimensional $SL(3)$ acting as

$$g \cdot x = \frac{ax + bu + c}{hx + ku + \ell}$$

$$g \cdot u = \frac{dx + eu + f}{hx + ku + \ell} \qquad (6.3)$$

where

$$g = \begin{pmatrix} a & b & c \\ d & e & f \\ h & k & \ell \end{pmatrix} \tag{6.4}$$

has determinant 1. It is straightforward to check both that

$$\widetilde{u_{xx}} = -\left(\frac{hx + ku + \ell}{ch - a\ell + (bh - ak)u + (ck - b\ell + (ak - bh)x)u_x} \right)^3 u_{xx}$$

and that the line $\alpha x + \beta u + \gamma = 0$ maps to another, $\widetilde{\alpha} x + \widetilde{\beta} u + \widetilde{\gamma} = 0$.

Exercise 6.1.3 In Example 6.1.2, calculate the induced action on the coefficients α, β and γ.

If $h(t)$ is a one parameter subgroup of G, then setting $g = h(t)$ into equation (6.2), differentiating with respect to t at $t = 0$, and evaluating the result on the surface $\Delta = 0$ in $(x, u^\alpha, u_K^\alpha)$-space, yields the *infinitesimal criterion* for $h(t)$ to be a symmetry of $\Delta = 0$, namely

$$(\mathbf{v}_h \Delta)|_{\Delta=0} = 0 \tag{6.5}$$

where

$$\mathbf{v}_h = \sum_{\alpha, K} \left(\xi \partial_x + \phi^\alpha \partial_{u^\alpha} + \phi_K^\alpha \partial_{u_K^\alpha} \right),$$

and where, from Definition 1.6.12,

$$\left. \frac{\partial \widetilde{x}}{\partial a_j} \right|_{g=e} = \xi_j, \qquad \left. \frac{\partial \widetilde{u}^\alpha}{\partial a_j} \right|_{g=e} = \phi_{,j}^\alpha, \qquad \left. \frac{\partial \widetilde{u_K^\alpha}}{\partial a_j} \right|_{g=e} = \phi_{K,j}^\alpha \tag{6.6}$$

are the infinitesimals of the group action. Recall the infinitesimal action on the derivatives u_K^α is obtained via the prolongation formula given in Section 1.6.

Lie's algorithm (see Remark 1.6.16) for finding the symmetry group of a differential system finds all possible $\mathbf{v}_h \cdot x = \xi(x, u)$ and $\mathbf{v}_h \cdot u^\alpha = \phi^\alpha(x, u)$, such that equation (6.5) holds. For systems, there are some subtleties involved in whether equation (6.5) implies Δ is a relative invariant system (Olver, 1993, Section 2.6).

Example 6.1.4 Consider the equation

$$\Delta : \qquad y_{xx} - \frac{2}{y} y_x^2 - \frac{1}{x} y_x - \frac{1}{x} y^2 = 0. \tag{6.7}$$

v	ξ	ϕ
$\mathbf{v}_1$	$\dfrac{1}{xy} - 1$	$-\dfrac{y}{x}$
$\mathbf{v}_2$	$-x^2 + \dfrac{x}{y}$	$3xy - 2$
$\mathbf{v}_3$	$-\dfrac{1}{x}$	$\dfrac{y^2}{x}$
$\mathbf{v}_4$	x	$-xy^2$
$\mathbf{v}_5$	x^3	$x^3y^2 - 2x^2y$
$\mathbf{v}_6$	0	$-xy^2 + y$
$\mathbf{v}_7$	0	x^2y^2
$\mathbf{v}_8$	0	y^2

Figure 6.2 The infinitesimals for the symmetry group of equation (6.7).

Lie's algorithm yields ξ and ϕ depending on eight different constants, and thus the symmetry group has dimension eight; the coefficient of each constant gives a different basis element for the Lie algebra. Figure 6.2 gives ξ and ϕ for each basis element. Taking the fourth symmetry vector $\mathbf{v}_4 = x\partial_x - xy^2\partial_y$, prolonging it to order 2 and applying it to Δ yields

$$[(x\partial_x - xy^2\partial_y - (y^2 + 2xyy_x + y_x)\partial_{y_x}$$
$$-2(2yy_x + xy_x^2 + xyy_{xx} + y_{xx})\partial_{y_{xx}}]\Delta$$
$$= -2(1 + xy)\Delta$$

as required. We leave it as an exercise for the reader to verify the other seven symmetry vectors. We note that not all software implementing Lie's algorithm finds all eight symmetry vectors for this example.

Remark 6.1.5 It is a theorem that if a second order ordinary differential equation $\Delta(x, y, y_x, y_{xx}) = 0$ has an eight dimensional symmetry group, then there exist coordinates $X = X(x, y)$, $Y = Y(x, y)$ such that the equation becomes $Y_{XX} = 0$. See for example Ibragimov (1992).

6.2 Solving invariant ordinary differential equations using moving frames

In this section we give a detailed outline of how to solve an invariant ordinary differential equation, given a symmetry (sub)group of dimension less than the order of the equation. In fact, we need also to assume that the prolonged action of the symmetry group is free and regular in some domain in $(x, u, \ldots, \mathrm{d}^n u/\mathrm{d}x^n)$ space where n is the order of the equation and where the domain includes solution curves to the equation.

Step 0

We are given an ordinary differential equation with symmetry group G. If the independent variable is not invariant under the action of G, we reparametrise solution curves to

$$x = x(s), \qquad u = u(s), \qquad u_x = \frac{u_s}{x_s}, \ldots$$

and set $\tilde{s} = s$. This ensures

$$\mathcal{D}_s = \frac{d}{ds}.$$

Thus the equation $\Delta = 0$ can be written in the form

$$F(x, x_s, \ldots, u, u_s, \ldots) = 0,$$

where we have used F instead of Δ to distinguish the reparametrised equation from the original. The variable s will not appear explicitly in this equation.

Exercise 6.2.1 Show the chain rule yields

$$\frac{d}{dx} = \frac{1}{x_s} \frac{d}{ds}$$

and hence that

$$\frac{dy}{dx} = \frac{y_s}{x_s}, \qquad \frac{d^2 y}{dx^2} = \frac{1}{x_s} \frac{d}{ds}\left(\frac{1}{x_s} \frac{dy}{ds}\right) = \frac{x_s y_{ss} - y_s x_{ss}}{x_s^3}.$$

Find $d^3 y/dx^3$ in terms of x, y and their derivatives. Note: in this context, the chain rule is also known as implicit differentiation.

The equation $F = 0$ is invariant under arbitrary reparametrisations, $\tilde{s} = f(s)$, and we may take a *companion equation*,

$$F_c(s, x, x_s, u, u_s, \ldots) = 0$$

in order to have a well-determined system. The companion equation can be thought of as fixing the parametrisation. It turns out that we do not actually need to solve it, thus we may choose it so as to ease the calculations provided the resulting system has the correct dimension of solution space. The companion equation needs to be compatible with the equation being solved and the choice of companion equation is justified at the end of the calculation. If the system $F = 0$, $F_c = 0$ has the same dimension solution space as $\Delta = 0$ we say the companion equation is compatible, otherwise not.

Step 1

We suppose now that $F = 0$ is of order at least that of the dimension of the group, and that $\eta = I_{K^0}^x$ and $\sigma = I_{J^0}^u$ are the generating invariants for some multi-indices K^0 and J^0, so that the equation takes the form,

$$0 = F(I^x, I_1^x, \ldots, I^u, I_1^u, \ldots) = \mathcal{F}(\eta, \eta_s, \ldots, \sigma, \sigma_s, \ldots).$$

Note that s will not appear explicitly. This rewrite uses Theorem 4.4.9, the Replacement Theorem, as well as the symbolic differentiation formulae to write higher order invariants in terms of η and σ and their derivatives. Similarly, the companion equation becomes

$$0 = \mathcal{F}_c(s, \eta, \eta_s, \ldots, \sigma, \sigma_s, \ldots).$$

If the frame is such that more than one invariant is needed to generate the set $\{I_L^u \,|\, |L| \geq 0\}$ or the set $\{I_L^x \,|\, |L| \geq 0\}$, then syzygies between the generating invariants need to be included in the differential system for the invariants.

If x is an invariant, then one does not need a companion equation, and η will not appear.

We thus obtain a system of ODEs $I(\Delta) = 0$, namely $\mathcal{F} = \mathcal{F}_c = 0$, for the generating invariants.

Step 2

Solving $I(\Delta) = 0$ for the generating invariants, we obtain the entries of the curvature matrix $\mathcal{Q}$ as functions of s.

Step 3

Next, we solve the equation $\varrho_s = \mathcal{Q}\varrho$, equation (5.2), for $\varrho(s)$, and from this we obtain the frame $\rho(s)$ in parameter form.

Step 4

Once we have ρ on solutions of the invariantised equation, we obtain x and u and indeed all the derivatives of u from

$$\begin{aligned}
x(s) &= \rho^{-1}(s) \cdot I^x(s) \\
u(s) &= \rho^{-1}(s) \cdot I^u(s) \\
u_s(s) &= \rho^{-1}(s) \cdot I_1^u(s) \\
u_{ss}(s) &= \rho^{-1}(s) \cdot I_{11}^u(s) \\
&\;\;\vdots
\end{aligned} \tag{6.8}$$

where the action, $\cdot$, is the one relevant for that jet coordinate. Note that the initial data for the $\rho_s = \mathcal{Q}\rho$ equation yield initial data for x, u, u_s and so on. A close look at Figure 6.1 reveals that some redundancy in the initial data required by the moving frame solution method will exist.

Discussion

We observe the following.

- We do not need to solve the companion ODE for x explicitly.
- We do not need to know what the I_K^x and I_K^u are as explicit expressions of the derivatives of u and x.
- To obtain $\mathcal{F} = 0$ from $\Delta = 0$ we need only the normalisation equations. To obtain $\mathcal{Q}$ we need know only the infinitesimal action and the normalisation equations.
- We can obtain x and u numerically from knowing ρ numerically.

We discuss the implications of our method for symbolic and numeric computation in more detail in Section 6.6.

6.3 First order ordinary differential equations

We illustrate the method outlined above on a simple first order equation. See Cantwell (2002), Chapter 6, for an excellent discussion of using symmetries to integrate first order ordinary differential equations. Here we show how moving frames and the method introduced in this chapter contribute to the topic.

Consider the ordinary differential equation

$$xy_x^2 - 3yy_x + 9x^2 = 0. \tag{6.9}$$

First order equations are equivalent to vector fields and the flow of the corresponding vector field will be a symmetry of the equation. Finding that particular symmetry action is equivalent to solving the equation, so one looks for an additional symmetry, either using look-up tables such as that in Cantwell (2002), page 158, or by inspection. Only a single one parameter group action is needed for a first order equation. If the action is a translation in either the independent or dependent variable, the ODE may be integrated immediately by quadratures. We discuss these cases at the end of this section for completeness.

For equation (6.9) a suitable symmetry group action is

$$\alpha \cdot x = \exp(2\alpha)x, \qquad \alpha \cdot y = \exp(3\alpha)y \tag{6.10}$$

so $\alpha \in G = (\mathbb{R}, +)$. Since x is not invariant, we reparametrise to $x = x(s)$, $y = y(s)$ and the equation becomes

$$x\left(\frac{y_s}{x_s}\right)^2 - 3y\frac{y_s}{x_s} + 9x^2 = 0. \tag{6.11}$$

Taking the normalisation equation to be $\widetilde{x} = 1$ (we leave it to the reader to work out the domain of the frame) the error matrix is

$$\mathbf{K} = \begin{pmatrix} \tfrac{1}{2} I_1^x \end{pmatrix},$$

the invariant operator is $\mathcal{D} = d/ds$, and we have $\mathcal{D} I^y = I_1^y - \tfrac{3}{2} I_1^x I^y$. Taking the companion equation to be $I_1^x = 1$, the invariantised equation is

$$\left(\frac{d}{ds} I^y \right)^2 - \frac{9}{4} (I^y)^2 + 9 = 0.$$

This is easily solved, as it is separable, and the general solution is

$$I^y = \frac{y_0}{2} \exp(\tfrac{3}{2} s) + \frac{2}{y_0} \exp(-\tfrac{3}{2} s)$$

where $y_0 \neq 0$ is a constant of integration. There are also two special solutions $I^y \equiv \pm 2$.

The matrix representation of the group is

$$\exp(\alpha) \rightarrow (\exp(\alpha)) \in GL(1).$$

Solving

$$\varrho_s = \mathbf{K}\varrho = (\tfrac{1}{2} I_1^x)\varrho = (\tfrac{1}{2})\varrho$$

gives $\varrho = (\rho_0 \exp(s/2))$ where ρ_0 is a constant. Hence looking at the general solution of the invariantised equation,

$$x(s) = \rho^{-1} \cdot I^x = \rho^{-1} \cdot 1 = \frac{1}{\rho_0^2} \exp(-s),$$

$$y(s) = \rho^{-1} \cdot I^y$$

$$= \frac{1}{\rho_0^3} \exp(-\tfrac{3}{2} s) \left(\frac{y_0}{2} \exp(\tfrac{3}{2} s) + \frac{2}{y_0} \exp(-\tfrac{3}{2} s) \right)$$

$$= \frac{2}{\rho_0^3 y_0} \exp(-3s) + \frac{y_0}{2\rho_0^3}.$$

Eliminating s, we have

$$y = kx^3 + \frac{1}{k}, \qquad k = \frac{2\rho_0^3}{y_0} \tag{6.12}$$

and it is easily checked that this is the general solution to equation (6.9). This family of cubics is plotted in Figure 6.3 for real x and y.

Obtaining the solutions to (6.9) corresponding to $I^y \equiv \pm 2$ yields $y = \pm 2x^{3/2}$, which form the envelope to the family of cubics, also plotted in Figure 6.3. This curve is mapped to itself under the group action; restricting to real

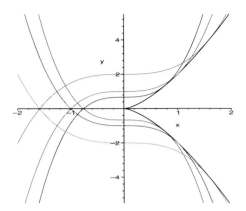

Figure 6.3 The solution curves for real x and y to equation (6.9).

x and y, it consists of three orbits of the action, since the origin is a fixed point.

Note that we needed only the normalisation equation, the infinitesimals and the explicit group action. We did not need to solve for the frame from the normalisation equations, nor did we need explicit expressions for I^y and I_1^x in terms of x and y, nor did we need to solve the companion equation $I_1^x = 1$ for x to obtain the (implied) parametrisation.

Exercise 6.3.1 Obtain the action induced on the solution curves by (6.10) given in equation (6.12) or, equivalently, find the induced action on the constants of integration.

Finally we note that if a first order ordinary differential equation $\Delta(x, u, u_x) = 0$ is invariant under the group action $\tilde{x} = x + \epsilon$, then it is of the form $u_x = f(u)$ which can be solved implicitly by quadratures,

$$\int \frac{\mathrm{d}u}{f(u)} = x + c$$

so moving frames are not required in this case. Similarly if $\Delta(x, u, u_x) = 0$ is invariant under $\tilde{u} = u + \epsilon$ then it is of the form $u_x = f(x)$, with solution $u(x) = \int^x f(t) \, \mathrm{d}t$.

Remark 6.3.2 The classical method of integrating an ODE invariant under a one parameter group action involves finding the coordinate system in which the group action is translation, the so-called canonical coordinates guaranteed by Frobenius' Theorem. One of the joys of the moving frame method is that the need to solve for the canonical coordinates is removed.

6.4 *SL(2)* invariant ordinary differential equations

In this section we consider ordinary differential equations invariant under the actions of $SL(2)$ listed in Example 1.2.14. For ODEs invariant under the first action, we give a moving frame proof of Schwarz' Theorem. An example of an ODE invariant under the third action is the Chazy equation, which is considered in Section 6.4.2.

6.4.1 Schwarz' Theorem

Consider the first of the $SL(2)$ actions listed in Example 1.2.14, but with the independent variable invariant, which is

$$\widetilde{x} = x, \qquad \widetilde{u} = \frac{au + b}{cu + d}, \qquad ad - bc = 1.$$

The generating invariant is

$$\{u; x\} = \frac{u_{xxx}}{u_x} - \frac{3}{2} \left(\frac{u_{xx}}{u_x} \right)^2 \tag{6.13}$$

also known as the Schwarzian derivative of u with respect to x.

Theorem 6.4.1 (Schwarz' Theorem) *The general solution of*

$$\{u; x\} = F(x),$$

where $\{u; x\}$ is defined in (6.13), is given by

$$u(x) = \frac{\psi_1(x)}{\psi_2(x)} \tag{6.14}$$

where the ψ_i are independent solutions of

$$\psi_{xx} + \frac{1}{2} F(x) \psi = 0. \tag{6.15}$$

The formula for the general solution can be verified directly. However, the use of moving frames allows one to *derive* it, and hence we can adapt the proof we give here to solve a multitude of similar problems.

Proof Since the independent variable is invariant, there is no need to reparametrise. Take the normalisation equations to be

$$\widetilde{u} = 0, \qquad \widetilde{u_x} = 1, \qquad \widetilde{u_{xx}} = 0,$$

and then in the standard matrix representation of $\mathfrak{sl}(2)$ we have,

$$Q = \begin{pmatrix} 0 & -1 \\ \frac{1}{2} I_{111}^u & 0 \end{pmatrix},$$

as was calculated in Example 5.2.6. Since the invariantised ordinary differential equation is

$$I_{111}^u = F(x)$$

we have that the equation for the frame is

$$\varrho_x = \begin{pmatrix} 0 & -1 \\ \dfrac{1}{2}F(x) & 0 \end{pmatrix} \varrho.$$

Setting

$$\varrho = \begin{pmatrix} \sigma_1 & \tau_1 \\ \sigma_2 & \tau_2 \end{pmatrix}$$

we obtain

$$\sigma_{1,x} = -\sigma_2, \qquad \sigma_{2,x} = \frac{1}{2}F(x)\sigma_1$$

so that

$$\sigma_{1,xx} + \frac{1}{2}F(x)\sigma_1 = 0$$

and similarly for the τ_i. That is, both σ_1 and τ_1 satisfy equation (6.15). Taking σ_1 and τ_1 to be two non-zero independent solutions of (6.15) with Wronskian $\sigma_{1,x}\tau_1 - \sigma_1\tau_{1,x}$ equal to one, we have that

$$\varrho^{-1} = \begin{pmatrix} -\tau_{1,x} & -\tau_1 \\ \sigma_{1,x} & \sigma_1 \end{pmatrix}.$$

Hence,

$$
\begin{aligned}
u(x) &= \rho(x)^{-1} \cdot I^u \\
&= \frac{-\tau_{1,x} I^u - \tau_1}{\sigma_{1,x} I^u + \sigma_1} \\
&= -\frac{\tau_1}{\sigma_1}
\end{aligned}
$$

as $I^u = 0$. $\square$

Exercise 6.4.2 Show that the Wronskian $\sigma_{1,x}\tau_1 - \sigma_1\tau_{1,x}$ is a constant when σ_1 and τ_1 both satisfy $\psi_{xx} + \frac{1}{2}F(x)\psi = 0$, and that it may be taken equal to one without loss of generality. Hint: consider the constants of integration $u(0)$, $u_x(0) \neq 0$ and $u_{xx}(0)$ in terms of $\sigma_1(0) \neq 0$, $\sigma_{1,x}(0)$ and $\tau_1(0)$.

Exercise 6.4.3 Use the method of moving frames to find the analogue of Schwarz' Theorem for

$$(\{u;x\})_x + \{u;x\} = F(x).$$

Hint: first solve $w_x + w = F$.

6.4.2 The Chazy equation

The Chazy equation is

$$u_{xxx} = 2uu_{xx} - 3u_x^2$$

and more generally,

$$u_{xxx} = 2uu_{xx} - 3u_x^2 + \alpha(6u_x - u^2)^2 \tag{6.16}$$

which arises in many contexts (Ablowitz and Clarkson, 1991). One form of the general solution was written down by Chazy, who gave no hint as to how he found it. Another form of the solution was obtained using Lie symmetry theory by Clarkson and Olver (1996). The form of the solution found by symmetry methods differs from that written down by Chazy. A straightforward application of moving frames yields the Clarkson and Olver form of the solution. It is not known whether Chazy's form of the solution can be found by a different choice of the normalisation equations.

The equation (6.16) is invariant under the action of $SL(2)$, given by

$$\widetilde{x} = \frac{ax + b}{cx + d}$$

$$\widetilde{u} = 6c(cx + d) + (cx + d)^2 u$$

where $ad - bc = 1$, which is the third of those listed in Example 1.2.14. The infinitesimal action in table form is

	x	u
a	$2x$	$-2u$
b	1	0
c	$-x^2$	$6 + 2xu$

Since the action does not leave x invariant, we reparametrise. Thus we set $x = x(s)$ and $u = u(s)$ with the same group action as above but adding now $\widetilde{s} = s$. The invariant differential operator is then d/ds. Taking the normalisation equations,

$$\widetilde{x} = 0, \qquad \widetilde{x}_s = 1, \qquad \widetilde{u} = 0$$

we have in the standard matrix representation of $SL(2)$ that

$$\frac{d}{ds}\varrho = -\begin{pmatrix} \frac{1}{2}I_{11}^x & 1 \\ \frac{1}{6}I_1^u & -\frac{1}{2}I_{11}^x \end{pmatrix}\varrho, \qquad (6.17)$$

which was calculated in Exercise 5.2.7. We take the companion ODE to be

$$I_{11}^x = 0,$$

a choice which will be vindicated by the calculations which follow; that is, we obtain a three dimensional set of solutions. The invariantisation of the reparametrisation of equation (6.16) is then

$$\frac{d^2}{ds^2}I_1^u + (4 - 36\alpha)(I_1^u)^2 = 0$$

showing that I_1^u is an elliptic function of s. Next, if $(\sigma_1, \sigma_2)^T$ is the first column of ϱ, then equation (6.17) implies that

$$\sigma_{1,ss} = \frac{1}{6}I_1^u\sigma_1.$$

A linear Schrödinger equation whose potential is an elliptic function is a Lamé equation (Whittaker and Watson, 1952). So, two independent solutions of this Lamé equation with Wronskian equal to one yield ρ. Using equations (6.8) then yields the solutions of the Chazy equation, without having solved for I_1^u or I_{11}^x in terms of $u(s)$ and $x(s)$.

Remark 6.4.4 To show equations (6.8) are consistent with the actual expression for ϱ given in equation (5.5) and calculated in Exercise 5.2.7, we note that the frame in matrix form is,

$$\varrho = \begin{pmatrix} \dfrac{1}{\sqrt{x_s}} & -\dfrac{x}{\sqrt{x_s}} \\ -\dfrac{1}{6}u\sqrt{x_s} & \dfrac{1}{6}(xu + 6)\sqrt{x_s} \end{pmatrix}. \qquad (6.18)$$

We then have

$$x(s) = \rho^{-1} \cdot I^x = \frac{\rho_{2,2}I^x - \rho_{1,2}}{-\rho_{2,1}I^x + \rho_{1,1}} = -\frac{\rho_{1,2}}{\rho_{1,1}} = x(s)$$

using $I^x = 0$ and

$$\begin{aligned} u(s) &= \rho^{-1} \cdot I^u \\ &= 6(-\rho_{2,1})(-\rho_{2,1}I^x + \rho_{1,1}) + (-\rho_{2,1}I^x + \rho_{1,1})^2 I^u \\ &= 6(-\rho_{2,1})(\rho_{1,1}) \\ &= u(s) \end{aligned}$$

since both $I^x = 0$ and $I^u = 0$ from the normalisation equations. The consistency is clear.

To show the necessity of reparametrisation, it is instructive to apply the moving frame reduction method to the Chazy equation in the original variables, $(x, u(x))$. Using the normalisation equations,

$$\widetilde{x} = 0, \qquad \widetilde{u} = 0, \qquad \widetilde{u}_x = 1$$

the frame in matrix form is,

$$\rho = \begin{pmatrix} -\left(u_x - \dfrac{u^2}{6}\right)^{1/4} & x\left(u_x - \dfrac{u^2}{6}\right)^{1/4} \\ \dfrac{u}{6}\left(u_x - u^2/6\right)^{-1/4} & -(1 + xu/6)\left(u_x - u^2/6\right)^{-1/4} \end{pmatrix}.$$

The invariant differential operator is

$$\mathcal{D} = \frac{1}{\left(u_x - u^2/6\right)^{1/2}} \frac{\mathrm{d}}{\mathrm{d}x}$$

and the generating invariant is

$$I_{11}^u = \frac{u_{xx} - uu_x + u^3}{(u_x - u^2/6)^{3/2}}.$$

Using the Replacement Theorem on equation (6.16) yields

$$I_{111}^u = 36\alpha - 3$$

and we also have,

$$\mathcal{D}I_{11}^u = I_{111}^u - \frac{3}{2}(I_{11}^u)^2 - 1.$$

Thus, we obtain the invariantised equation,

$$\mathcal{D}I_{11}^u + \frac{3}{2}(I_{11}^u)^2 - 36\alpha + 4 = 0$$

which superficially has the form of a Ricatti equation in terms of this particular generating invariant. Unfortunately, the coefficient in the invariant differential operator renders the equation highly non-trivial to solve.

6.5 Equations with solvable symmetry groups

It is a classical theorem that an ordinary differential equation of order n, with a solvable symmetry group of dimension n, can be integrated by quadratures. Here we show how moving frames greatly simplifies the calculations involved.

The classical method involves reducing the ODE by one group parameter at a time, and the order of reduction is the same as that used when computing the frame as a product in Section 4.6. However, using the moving frame solution method one parameter at a time is *not* the route recommended here. A better way is to use the fact that a solvable Lie algebra has a faithful representation using upper triangular matrices. Then the curvature matrix will be upper triangular, leading to a triangular system to solve for ρ.

We show what happens by integrating equation (4.22),

$$y_{xx} + \frac{1}{x}y_x + e^y = 0$$

whose solvable symmetry group was found in Exercise 4.6.4; we repeat it here for convenience. The group is $G = \mathbb{R}^+ \ltimes \mathbb{R}^+$, the action is

$$\widetilde{x} = ax^b, \qquad \widetilde{y} = y - 2\log(ab) - 2(b-1)\log x$$

with identity element $(b, a) = (1, 1)$, and an upper triangular matrix representation of this group is given by

$$(b, a) \rightarrow \begin{pmatrix} b & \log a \\ 0 & 1 \end{pmatrix}. \tag{6.19}$$

Using the moving frame reduction method, finding the exact solution of this non-linear equation involves three straightforward integration steps. Since the independent variable is not invariant, we reparametrise to $x = x(s)$, $u = u(s)$ and set $\widetilde{s} = s$. If we take the normalisation equations to be $\widetilde{x} = 1$, $\widetilde{y} = 0$, we have that

$$\mathsf{K} = s \ \overset{\displaystyle a \qquad \quad b}{\left(-I_1^x \quad I_1^x + \tfrac{1}{2}I_1^y \right).}$$

Taking the companion equation to be $I_1^x = 1$, we have from $\mathrm{d}/\mathrm{d}s\, I_1^x = 0$ that $I_{11}^x = -\tfrac{1}{2}I_1^y$, and also

$$\frac{\mathrm{d}}{\mathrm{d}s}I_1^y = I_{11}^y - 2 - I_1^y$$

and hence the invariantisation of the reparametrised equation becomes

$$\frac{\mathrm{d}}{\mathrm{d}s}I_1^y + \tfrac{1}{2}(I_1^y)^2 + 2I_1^y + 3 = 0.$$

Thus

$$I_1^y(s) = \sqrt{2}\tan\left(\frac{s_0 - s}{\sqrt{2}} \right) - 2.$$

Using the faithful matrix representation of the group given in equation (6.19) above, a basis for the Lie algebra needed to calculate $\mathcal{Q}$ in the equation for ϱ

is, using equation (5.3),

$$\mathfrak{a}_b = \begin{pmatrix} 1 & 0 \\ 0 & 0 \end{pmatrix}, \qquad \mathfrak{a}_a = \begin{pmatrix} 0 & 1 \\ 0 & 0 \end{pmatrix},$$

and thus using Theorem 5.2.4, the equation for ϱ becomes, with $I_1^x = 1$,

$$\begin{pmatrix} b_s & (\log a)_s \\ 0 & 0 \end{pmatrix} = \begin{pmatrix} 1 + \frac{1}{2}I_1^y & -1 \\ 0 & 0 \end{pmatrix} \begin{pmatrix} b & \log a \\ 0 & 1 \end{pmatrix}.$$

It can be seen that the equations for a and b separate. With the value of I_1^y obtained above, we have

$$b(s) = b_0 \cos\left(\frac{s_0 - s}{\sqrt{2}}\right)$$

and

$$\log a(s) = \cos\left(\frac{s_0 - s}{\sqrt{2}}\right)\left[a_0 + \sqrt{2}\log\left(\sec\left(\frac{s_0 - s}{\sqrt{2}}\right) + \tan\left(\frac{s_0 - s}{\sqrt{2}}\right)\right)\right].$$

From the group multiplication law and group action formulae given in Exercise 4.6.4 we have

$$(b, a)^{-1} = (b^{-1}, a^{-1/b}),$$

and thus

$$x(s) = \rho(s)^{-1} \cdot I^x = a(s)^{-1/b(s)}(I^x)^{1/b(s)} = \exp(-(\log a(s))/b(s))$$

and

$$y(s) = \rho(s)^{-1} \cdot I^y = I^y - \log\left(b(s)^{-1}a(s)^{-1/b(s)}\right) - 2(b^{-1} - 1)\log I^x$$
$$= 2\log b(s) + 2(\log a(s))/b(s).$$

That this is a solution of equation (4.22) above for x positive can be verified by direct substitution. Solutions for x negative can be obtained by noting that the equation is invariant under $(x, y) \mapsto (-x, y)$. We note that we appear to have three constants of integration, s_0, b_0 and f_0, for a second order equation. The third has been introduced by the freedom in setting where on your solution curve you take $s = 0$. However, it should be noted that in fact there are three values needed to determine the two initial points (x_0, y_0) and $(x_0, y_x(x_0))$, namely x_0, y_0 and $y_x(x_0)$.

6.6 Notes on symbolic and numeric computation

Here we summarise what is known that would allow the method to be implemented in a symbolic and numerical computation environment.

(i) The reduced equation is obtained via the Replacement Theorem 4.4.9, which needs only the normalisation equations.

(ii) Given the infinitesimal vector fields, a faithful matrix representation of the finite dimensional part of the Lie algebra can be obtained by Ado's Theorem. This has been implemented by de Graaf (2000). However, in practice the matrices obtained can be unwieldy.

(iii) By Theorem 5.2.4 the matrix

$$A = \rho_s \rho^{-1}$$

can be calculated knowing only the normalisation equations and the infinitesimal action, in the faithful representation at hand.

(iv) Equations of the form

$$\varrho_s = A(s)\varrho$$

with $A(s) \in \mathfrak{g}$ where $\mathfrak{g}$ is a matrix Lie Algebra can be solved numerically keeping $\rho \in G$ using geometric integrators, in particular the 'Magnus' or 'Lie group integrators' (Iserles *et al.*, 2000). There is the possibility of choosing both the normalisation equations, the companion ODE and the representation to adapt the 'shape' of A to ease the integration.

(v) Applying equations (6.8) requires knowing the group action, not just the infinitesimals. If you know only the infinitesimal action, which is how symmetry groups of DEs are given by symbolic software, then the group action can be obtained by integration of the system (1.54), see Theorem 1.6.23.

6.7 Using only the infinitesimal vector fields

In this section we give the vector field counterpart to Theorem 5.2.4, and apply it to calculation of the moving frame reduction solution. The calculation of a faithful matrix representation of the Lie algebra can thus be avoided, but the resulting equation for ρ may be, conceptually at least, significantly harder to solve.

Theorem 6.7.1 *Suppose $G \times M \to M$ is a smooth left group action, $(a_1, \ldots, a_r)$ are local coordinates on G, $(z_1, \ldots, z_n)$ are local coordinates* *on M and*

$$\zeta_j^i(z) = \left.\frac{\partial g \cdot z_i}{\partial a_j}\right|_{g=e}$$

are the infinitesimals of the action with respect to these coordinates. If $h(s)$ is a path in G and $h(s) \cdot z_i = \widetilde{z}_i$, then

$$\frac{d}{ds}\widetilde{z}_i = \sum_\ell \left(T_e R_h^{-1}\frac{d}{ds}h\right)_\ell \zeta_\ell^i(\widetilde{z}_i). \tag{6.20}$$

If $h(s) = \rho(z(s))$ where ρ is a right moving frame for the action, then

$$\frac{d}{ds}\widetilde{z}_i = \sum_\ell \mathbf{K}_{s\ell}\zeta_\ell^i(\widetilde{z}_i(s)) \tag{6.21}$$

where $\mathbf{K}$ is given in Theorem 4.5.5.

Proof The starting point is Theorem 3.2.37 which we now briefly recall: given $(a_1, a_2, \ldots, a_r)$ coordinates in some domain in G, and letting $h(s) \in G$ be a path in that chart domain, then for each coordinate z_i of $z \in M$,

$$\left.\begin{pmatrix} \dfrac{\partial g \cdot z_i}{\partial a_1} \\ \vdots \\ \dfrac{\partial g \cdot z_i}{\partial a_r} \end{pmatrix}\right|_{g=h} = (T_e R_h)^{-T} \begin{pmatrix} \zeta_1^i(h \cdot z) \\ \vdots \\ \zeta_r^i(h \cdot z) \end{pmatrix}. \tag{6.22}$$

Since the argument is the same for each coordinate of z, we drop the subscript. Setting $h(s) \cdot z = \widetilde{z}(s)$ and $h(s) = (a_1(s), \ldots, a_r(s))$,

$$\begin{aligned}
\frac{d}{ds}\widetilde{z}(s) &= \sum_k a_k'(s)\frac{\partial}{\partial a_k}(\widetilde{x}(s)) \\
&= \sum_{k,\ell} a_k'(s)\left(T_e R_{h(s)}^{-T}\right)_{k\ell}(\zeta_\ell(\widetilde{z}(s)) \\
&= \sum_{k,\ell} \left(\left(T_e R\right)_{h(s)}^{-1}\right)_{\ell k} a_k'(s)\right) \zeta_\ell(\widetilde{z}(s)) \\
&= \sum_\ell \left(T_e R_{h(s)}^{-1}\frac{d}{ds}h(s)\right)_\ell \zeta_\ell(\widetilde{z}(s)).
\end{aligned} \tag{6.23}$$

This proves equation (6.20). Finally, to obtain equation (6.21), we set $h(s) = \rho(z(s))$ into equation (6.20), and use the rightmost formula for $\mathbf{K}$ in equation (4.13). $\qquad\square$

Equation (6.21) is essentially the differential equation for the frame ρ, not in the matrix form $\varrho_s = \mathcal{Q}\varrho$, but with ρ viewed as an element of a transformation group. While the matrix equation has $r = \dim(G)$ constants of integration, equation (6.21) has initial data $\rho(0) \cdot z_i$. Thus, it is still necessary to solve the ordinary differential system giving the general group action.

Applying Theorem 6.7.1 to the coordinates x and u, whose infinitesimals with respect to a_ℓ are denoted by ξ_ℓ and ϕ_ℓ respectively, we obtain

$$\frac{d}{ds}\widetilde{x}(s) = \sum_\ell \left(T_e R_\rho^{-1} \frac{d}{ds}\rho \right)_\ell \xi_\ell(\widetilde{x}(s))$$

$$= \sum_\ell \mathbf{K}_{s\ell} \xi_\ell(\widetilde{x}(s)) \tag{6.24}$$

$$\frac{d}{ds}\widetilde{u}(s) = \sum_\ell \mathbf{K}_{s\ell} \phi_\ell(\widetilde{u}(s)).$$

For the application we consider in this chapter, namely the solution of Lie group invariant ODEs, we know how the matrix $\mathbf{K}$ varies as a function of the invariant independent coordinate s once we have solved the invariantised ODE system. One then uses equations (6.24) to obtain the frame $\rho(x(s), u(s), \dots)$, and the ODE system in Theorem 1.6.24 for each group parameter to obtain the general group action. Finally, one can write down the general solution to the original invariant ODE system.

Example 6.7.2 As ever we consider the $SL(2)$ action

$$\widetilde{x} = x, \qquad \widetilde{u} = \frac{au + b}{cu + d}, \qquad ad - bc = 1.$$

Since x is invariant, we can take $s = x$. The infinitesimals are

$$\phi_a(x, u) = 2u, \qquad \phi_b(x, u) = 1, \qquad \phi_c(x, u) = -u^2.$$

Suppose that we know only the infinitesimals and not the group action, that we do not know a faithful matrix representation of the Lie algebra, and we have forgotten Schwarz' Theorem. The generic one parameter group action is the solution of the ordinary differential equation

$$\dot{u} = \alpha(2u) + \beta + \gamma(-u^2) \tag{6.25}$$

where α, β and γ are arbitrary constants, using Theorem 1.6.24 for each independent group parameter. Since we can see that ϕ_a is the infinitesimal of a scaling action and ϕ_b the infinitesimal of a translation, we can guess that $\widetilde{u} = 0$, $\widetilde{u}_x = 1$ are appropriate normalisation equations. Adding $\widetilde{u}_{xx} = 0$ yields

$$\mathbf{K} = x \begin{array}{ccc} a & b & c \\ \left(0 & -1 & \frac{1}{2}I_{111}^u \right) \end{array}$$

which can be obtained using only the infinitesimals and the normalisation equations. Applying equation (6.24) we obtain

$$\frac{\mathrm{d}}{\mathrm{d}x}\widetilde{u} = \begin{pmatrix} a & b & c \\ 0 & -1 & \frac{1}{2}I^u_{111} \end{pmatrix} \begin{pmatrix} 2\widetilde{u} \\ 1 \\ -\widetilde{u}^2 \end{pmatrix} \begin{matrix} a \\ b \\ c \end{matrix}$$

or

$$\frac{\mathrm{d}}{\mathrm{d}x}\widetilde{u} = -1 - \tfrac{1}{2}I^u_{111}\widetilde{u}^2. \tag{6.26}$$

Suppose we wish to solve the ODE, $\{u; x\} = 2\exp(x)$, which invariantises to

$$I^u_{111} = 2\exp(x). \tag{6.27}$$

Setting equation (6.27) into equation (6.26) with initial condition $\widetilde{u}(x) = u$, and solving, yields an expression $\widetilde{u}(x) = F(x, u) = \rho(x) \cdot u$ in terms of Bessel functions. Solving this expression for u (which equals $\rho^{-1}(x) \cdot \widetilde{u}(x)$) and setting $\widetilde{u}(x) = I^u(x)$ gives the desired solution. Since $I^u(x) \equiv 0$ we obtain

$$u = \frac{Y_0(2)J_0(2\exp(x/2)) - J_0(2)Y_0(2\exp(x/2))}{Y_1(2)J_0(2\exp(x/2)) - J_1(2)Y_0(2\exp(x/2))}.$$

This is indeed a solution of $\{u; x\} = 2\exp(x)$ as can be verified directly. The general solution is $(au + b)/(cu + d)$; obtaining this requires one to solve for the group action, equation (6.25) above.

7

Variational problems with symmetry

The Calculus of Variations is a classical subject with major applications in physics and engineering, and a long history of development. The variational method converts the problem of finding a curve or surface that minimises or maximises an integral functional, such as

$$(x, u(x)) \mapsto \int_a^b L(x, u, u_x, \dots) \, \mathrm{d}x$$

in the case of curves in the plane, to the problem of solving a differential equation, called the Euler–Lagrange equation.

For many problems, there is a Lie group invariance to the integral that arises naturally from the physical model being considered. The aim of this chapter is to use the calculus of invariants developed in Chapters 4 and 5 to gain insight into the role the symmetry plays in the calculation and analysis of the solution set of the Euler–Lagrange equation.

The introductory part of this chapter puts in one place everything we will need for the applications of moving frames that we consider. Recent accessible texts include van Brunt (2004) and MacCluer (2005), but older ones are still 'gold', such as Courant and Hilbert (1953).

7.1 Introduction to the Calculus of Variations

In this brief introductory section on the Calculus of Variations we collect together the computational methods and results needed to study variational problems with symmetry.

In the first set of problems we consider, it is desired to find some curve $(x, u(x))$ that minimises or maximises an integral,

$$(x, u(x)) \mapsto \mathcal{L}[u] = \int_a^b L(x, u, u_x, u_{xx}, \dots) \, \mathrm{d}x$$

on some interval $[a, b] \subset \mathbb{R}$. The integrand $L(x, u, u_x, \dots) \, \mathrm{d}x$ is called the *Lagrangian*. We say $L \, \mathrm{d}x$ is one dimensional since the integral is with respect to the single coordinate x, and that $L \, \mathrm{d}x$ is of order n if the highest derivative of u appearing in L is $\mathrm{d}^n u / \mathrm{d}x^n$ (but see Exercise 7.1.3).

Remark 7.1.1 It is important to note that it is the whole integrand, $L \, \mathrm{d}x$, not just the function L, which gives the variational problem $\int L \, \mathrm{d}x$. Considering only the function can lead to serious errors in both computation and theory.

Blanket assumption We will assume our Lagrangians to be smooth functions of a finite number of arguments, and that the functional analytic properties of the curves on which we evaluate $\mathcal{L}[u]$ are sufficient to ensure the calculations that follow are valid. In particular, we assume that we can interchange integration and differentiation, and that the extremal curves we seek have a sufficient number of smooth derivatives. An accessible text which considers the most important functional analytic niceties in the context of applications is MacCluer (2005).

If we suppose that the particular curve $(x, u(x))$ extremises $\mathcal{L}[u]$, then for a variation $u(x) \mapsto u(x) + \epsilon v(x)$ we will have

$$
\begin{aligned}
0 &= \frac{\mathrm{d}}{\mathrm{d}\epsilon}\Big|_{\epsilon=0} \mathcal{L}[u + \epsilon v] \\
&= \frac{\mathrm{d}}{\mathrm{d}\epsilon}\Big|_{\epsilon=0} \int_a^b L(x, u + \epsilon v, u_x + \epsilon v_x, u_{xx} + \epsilon v_{xx}, \dots) \, \mathrm{d}x \\
&= \int_a^b \left(\frac{\partial L}{\partial u} v + \frac{\partial L}{\partial u_x} v_x + \frac{\partial L}{\partial u_{xx}} v_{xx} + \cdots \right) \mathrm{d}x \\
&= \int_a^b \left[\left(\frac{\partial L}{\partial u} - \frac{\mathrm{d}}{\mathrm{d}x} \frac{\partial L}{\partial u_x} + \frac{\mathrm{d}^2}{\mathrm{d}x^2} \frac{\partial L}{\partial u_{xx}} + \cdots \right) v \right. \\
&\qquad \left. + \frac{\mathrm{d}}{\mathrm{d}x} \left(\frac{\partial L}{\partial u_x} v + \frac{\partial L}{\partial u_{xx}} v_{xx} - \left(\frac{\mathrm{d}}{\mathrm{d}x} \frac{\partial L}{\partial u_{xx}} \right) v + \cdots \right) \right] \mathrm{d}x \\
&= \int E(L) v \, \mathrm{d}x + \left[\frac{\partial L}{\partial u_x} v + \cdots \right]_a^b
\end{aligned}
\tag{7.1}
$$

where $\mathrm{d}/\mathrm{d}x$ is a total derivative operator. We have assumed that we may interchange integration and differentiation, and then used integration by parts. The last line defines the *Euler–Lagrange operator E* for one dimensional Lagrangians, namely

$$
E(L) = \sum_k (-1)^k \frac{\mathrm{d}^k}{\mathrm{d}x^k} \frac{\partial L}{\partial u_k}, \qquad u_k = \frac{\mathrm{d}^k}{\mathrm{d}x^k} u.
\tag{7.2}
$$

Since equation (7.1) holds for any variation v, we may take a variation that satisfies $v(a) = v(b) = v_x(a) = \cdots = 0$ and obtain that $\int_a^b E(L)v\,dx = 0$ for all variations with support interior to the domain $[a, b]$. It is then a result, called the Fundamental Lemma of the Calculus of Variations, that $E(L)$ is identically zero for all $x \in (a, b)$. In other words, a necessary condition for a curve to extremise $\mathcal{L}[u]$ is the differential equation $E(L) = 0$. Curves satisfying $E(L) = 0$ are said to be extremal curves for the variational problem $\int L\,dx$.

The calculation of the Euler–Lagrange equation is thus in two parts:

• the calculation of

$$\frac{d}{d\epsilon}\bigg|_{\epsilon=0} \mathcal{L}[u + \epsilon v],$$

• integration by parts.

The boundary or endpoint terms in (7.1) that accumulate as we perform the integration by parts play a vital role when considering problems without specific boundary conditions; if boundary conditions for the extremal curves are not specified by the application, then the so-called *natural* boundary conditions will apply. For example, if $u(a)$ is unspecified then $v(a)$ is free, and so the coefficient of $v(a)$ in the endpoint terms must be zero on an extremal; in the case of a first order Lagrangian, this is $\partial L/\partial u_x|_{x=a} = 0$, which is a condition on u and its derivatives at $x = a$.

The endpoint terms will also play a vital role in deriving Noether's Theorem in both the classical and invariantised cases, see Section 7.2.

In general, Euler–Lagrange equations are seriously non-linear and it is important to be able to investigate the solution set graphically, see for example Exercise 7.1.2.

Exercise 7.1.2 Show the Euler–Lagrange equation for

$$\mathcal{L}[u] = \int_0^{10} uu_x^2 + \sin(x)u^2 \, dx$$

is $u_x^2 - 2u \sin x + 2uu_{xx} = 0$. Show there exists a solution satisfying $u(0) = 1$, $u(10) = 4$. Hint: plot the solutions satisfying $u(0) = 1$ but with varying $u_x(0)$, and use a 'shooting argument', that is apply the Intermediate Value Theorem to the function $y \mapsto u^{[y]}(10)$ where $u^{[y]}$ is a solution satisfying $u_x^{[y]}(0) = y$. Is there more than one solution satisfying these boundary conditions?

Exercise 7.1.3 Show that if a Lagrangian $L\,dx$ satisfies

$$L(x, u, u_x, u_{xx}) = \frac{d}{dx}W(x, u, u_x) = \frac{\partial}{\partial x}W + u_x\frac{\partial}{\partial u}W + u_{xx}\frac{\partial}{\partial u_x}W,$$

then $E(L)$ is identically zero. Generalise the result to arbitrary order Lagrangians. Explain. Hint: can the integral $\int L \, dx$ change with variations to u inside the domain? Hence show that if two Lagrangians L_1 and L_2 differ by a total derivative, then their Euler–Lagrange equations will be equal. In this case we say L_1 and L_2 are *equivalent*; show this is indeed an equivalence relation. Formulate a definition of the order of a Lagrangian that is well defined on equivalence classes.

Exercises 7.1.4 and 7.1.7 give simple examples of how Euler–Lagrange equations behave under changes of variable. The general result can be seen in van Brunt (2004), Section 2.5, or Olver (1993), Theorem 4.8.

Exercise 7.1.4

(i) If $L = L(f(u), f'(u)u_x)$, show that

$$E^{f(u)}(L) = E^u(L)f'(u),$$

where $E^{f(u)}(L)$ means the Euler–Lagrange equation obtained by varying $f(u)$.

(ii) If $L = L(u_x, u_{xx})$ and $w = u_x$, so that $L = L(w, w_x)$, show that

$$E^u(L) = -\frac{d}{dx}E^w(L).$$

More generally,[†] show that if $w = A(u)$ where A is a linear differential operator, then

$$E^u(L) = A^* E^w(L),$$

where A^* is the *adjoint* of A; see Definition 7.3.2.

If the Lagrangian depends on more than one dependent variable, then an Euler–Lagrange system is obtained, one equation for each dependent variable.

Exercise 7.1.5 Generalise the above calculation in equation (7.1) to obtain the Euler–Lagrange system for a Lagrangian depending on more than one dependent variable, say u^α, $\alpha = 1, \ldots, q$ so that

$$L \, dx = L(x, u^\alpha, u^\alpha_x, \ldots) \, dx,$$

namely,

$$E^\alpha(L) = \sum_k (-1)^k \frac{d^k}{dx^k} \frac{\partial L}{\partial u^\alpha_k}, \qquad u^\alpha_k = \frac{d^k}{dx^k} u^\alpha, \qquad \alpha = 1, \ldots, q. \quad (7.3)$$

[†] This generalisation was pointed out to me by Peter Hydon.

Hint: calculate

$$\frac{\mathrm{d}}{\mathrm{d}\epsilon}\Big|_{\epsilon=0} \mathcal{L}[u^\alpha + \epsilon v^\alpha]$$

and note the v^α are independent variations.

Multivariable variational problems arise in the study of mechanical systems. Exercise 7.1.6 is the classic example.

Exercise 7.1.6 Even after all these years, still the best illustrative multivariable example is the *two body problem* with Newtonian gravity, also known as *Kepler's problem*. If the earth is at $(x^e(t), y^e(t), z^e(t))$ and the sun is at $(x^s(t), y^s(t), z^s(t))$, then setting $x = x^e - x^s$, $y = y^e - y^s$ and $z = z^e - z^s$, the Lagrangian is $L\,\mathrm{d}t$ where

$$L = \tfrac{1}{2}\left(x_t^2 + y_t^2 + z_t^2\right) - \frac{k}{(x^2 + y^2 + z^2)^{1/2}},$$

where k is a constant. Show that

$$E^x(L) = -x_{tt} + \frac{kx}{(x^2 + y^2 + z^2)^{3/2}},$$

and similarly for $E^y(L)$ and $E^z(L)$. First integrals of this system will be constructed in Exercise 7.2.7.

The point of the next exercise, 7.1.7, is to show that information is neither lost nor gained by reparametrisation of the Lagrangian.

Exercise 7.1.7 Consider a Lagrangian $L(x, u, u_x, u_{xx})\,\mathrm{d}x$ and change the parametrisation of the sought curve from $(x, u(x))$ to $(t, x(t), u(t))$; noting that $\mathrm{d}x = x_t\,\mathrm{d}t$ and

$$\frac{\mathrm{d}}{\mathrm{d}x} = \frac{1}{x_t}\frac{\mathrm{d}}{\mathrm{d}t},$$

we obtain

$$\mathcal{L}[u] = \int L\,\mathrm{d}x = \widetilde{\mathcal{L}}[x, u] = \int L\left(x, u, \frac{u_t}{x_t}, \frac{1}{x_t}\frac{\mathrm{d}}{\mathrm{d}t}\frac{u_t}{x_t}\right) x_t\,\mathrm{d}t = \int \widetilde{L}\,\mathrm{d}t,$$

where the last equality defines $\widetilde{L}$. Show that

$$x_t E^x(\widetilde{L}) + u_t E^u(\widetilde{L}) = 0, \tag{7.4}$$

where $\widetilde{L}$ is considered as having two dependent variables x and u both depending on t. Show further that

$$E^{u(t)}(\widetilde{L}) = x_t E^{u(x)}(L)$$

where the differential operators used to calculate $E^{u(t)}$ are powers of d/dt, and the differential operators used to calculate $E^{u(x)}$ are powers of d/dx. Note: equation (7.4) is a simple example of Noether's Second Theorem; for Lagrangians invariant under an infinite dimensional group, in this case $\tilde{t} = \tau(t)$ for an arbitrary smooth function τ, the Euler–Lagrange equations satisfy a differential identity.

Next, we need to consider variational problems with constraints. The main results are as follows. If you seek a curve that minimises $\mathcal{L}[u] = \int L \, dx$ amongst those curves that satisfy also $\int G(x, u, u_x, \dots) \, dx = A$ where A is a constant, then a necessary condition on the minimising curve $(x, u(x))$ is that $u(x)$ satisfies the system

$$E^u(\lambda_0 L - \lambda G) = 0, \qquad \int G \, dx = A \qquad (7.5)$$

where $\lambda_0 \in \{0, 1\}$, the constant λ is called the *Lagrange multiplier*, and λ_0 and λ are not both zero. Problems where $\lambda_0 = 0$ are called *abnormal*; they occur when the extremising curves also extremise $\int G \, dx$. Such extremal curves are called *rigid extremals*, see van Brunt (2004), Section 4.2.

The new parameter λ, which is free in the differential equation, is evaluated by calculating the integral constraint.

Exercise 7.1.8 Find the extremal of

$$\mathcal{L}[u] = \int_0^\pi u_{xx}^2 \, dx$$

subject to $u(0) = u_{xx}(0) = 0$, $u(\pi) = u_{xx}(\pi) = 0$ and

$$\int_0^\pi u^2 \, dx = 1.$$

Hint: write the equations for the constants of integration in matrix form, and consider the determinant of the matrix.

If the constraint is not an integral constraint, but must hold at each point of the curve, so that $G(x, u, u_x, \dots) = 0$ for all x, then the result is as follows: the necessary condition on the curve is that it satisfies the system

$$E^u(\lambda_0 L - \lambda(x)G) = 0, \qquad G = 0 \qquad (7.6)$$

where $\lambda_0 \in \{0, 1\}$, the function $\lambda(x)$ is now a function also called the Lagrange multiplier, and λ_0 and λ are not both identically zero. See van Brunt (2004), Section 6.2, for a discussion of the abnormal case where $\lambda_0 = 0$.

Remark 7.1.9 The function λ may, in principle, be eliminated from the system (7.6) using standard techniques from differential algebra to produce the differential equation for u, but this will produce Euler–Lagrange equations that are too high order. *Any constants of integration that can be absorbed into the Lagrange multiplier $\lambda(x)$ should be discarded as artefacts.* This is because the curve not only satisfies $G = 0$ but also $cG = 0$ for any constant c, and so forth.

The next exercise shows the need to absorb constants of integration for λ to obtain the correct Euler–Lagrange equations.

Exercise 7.1.10 Suppose a second order Lagrangian $L(x, u, u_x, u_{xx})\, dx$ is given, and that we wish to reparametrise the problem to $(s, x(s), u(s))$ where s is the Euclidean arc length, that is, where $x_s^2 + u_s^2 = 1$, see Figure 7.1. Show that the Euler–Lagrange equation $E^{u(x)}(L) = 0$ for $L\, dx$ is equivalent to the equation obtained by eliminating $\lambda(s)$ from the system

$$E^{x(s)}(\widetilde{L}) = 0, \qquad E^{u(s)}(\widetilde{L}) = 0, \qquad u_s^2 + x_s^2 = 1,$$

where

$$\widetilde{L}\, ds = \left[L(x, u, u_s/x_s, u_{ss}/x_s^4)x_s - \lambda(s)(x_s^2 + u_s^2 - 1) \right] ds.$$

The Lagrangian $\widetilde{L}\, ds$ is obtained by inserting into $L\, dx$ not only $dx = x_s\, ds$ but both $u_x = u_s/x_s$ and

$$u_{xx} = \left(\frac{1}{x_s} \frac{d}{ds} \right)^2 u$$

and then simplifying with respect to the constraint $x_s^2 + u_s^2 = 1$, and its differential consequence $x_s x_{ss} + u_s u_{ss} = 0$. Hint: calculating $0 = x_s E^{x(s)}(\widetilde{L}) + u_s E^{u(s)}(\widetilde{L})$ (see Exercise 7.1.7) yields

$$0 = 2\lambda_s + \frac{d}{ds}\left(u_s \frac{d}{ds}\left(\frac{D_4(L)}{x_s^3} \right) + 3D_4(L)\frac{u_{ss}}{x_s^3} \right),$$

where $D_4(L)$ is the derivative of L with respect to its fourth argument. The constant of integration can be absorbed into λ. Putting λ into $E^{x(s)}(\widetilde{L}) = 0$ yields $-u_s E^{u(x)}(L) = 0$ after simplifying with respect to $u_s^2 + x_s^2 = 1$ and its differential consequences.

7.1.1 Results and non-results for Lagrangians involving curvature

There are an amazing number of fallacious arguments and results appearing in the literature, including textbooks published by reputable mathematical

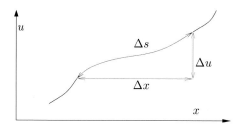

Figure 7.1 If s is Euclidean arc length, $x_s^2 + u_s^2 = 1$.

publishers, concerning Lagrangians involving Euclidean curvature and Euclidean arc length. Since we will be considering these Lagrangians in some depth, we take a moment to sort out the false from the true.

We recall that the Euclidean curvature of a curve $(x, u(x))$ is

$$\kappa = \frac{u_{xx}}{(1 + u_x^2)^{3/2}}$$

and in terms of a parametrised curve $(x(t), u(t))$ it is

$$\kappa = \frac{u_{tt}x_t - x_{tt}u_t}{(x_t^2 + u_t^2)^{3/2}}. \tag{7.7}$$

Lagrangians involving curvature typically set the curve parameter to be s which is arc length; this is the parameter for which

$$x_s^2 + u_s^2 = 1, \tag{7.8}$$

see Figure 7.1. In this case, we have

$$\kappa = u_{ss}x_s - x_{ss}u_s = u_{ss}/x_s = -x_{ss}/u_s$$

since $u_s u_{ss} + x_s x_{ss} = 0$. Most importantly, we have

$$ds = \sqrt{1 + u_x^2}\, dx \tag{7.9}$$

and equivalently,

$$\frac{d}{ds} = \frac{1}{\sqrt{1 + u_x^2}} \frac{d}{dx}. \tag{7.10}$$

The most famous Lagrangian involving curvature is

$$\mathcal{L}[u] = \int \kappa^2\, ds = \int \frac{u_{xx}^2}{(1 + u_x^2)^{5/2}}\, dx. \tag{7.11}$$

Since the Lagrangian involves u_{xx} and is not a total derivative, the Euler–Lagrange equation will be of order 4 in u (see Exercise 7.1.3), and hence will involve κ_{ss}; in particular, it is not $\kappa = c$. Even though straight lines, that is, curves which satisfy $\kappa = 0$, are indeed solutions, they are not the only ones.

Theorem 7.1.11 (Euler) *The Euler–Lagrange equation for the variational problem given in equation (7.11) is*

$$\kappa_{ss} + \tfrac{1}{2}\kappa^3 = 0. \tag{7.12}$$

Remark 7.1.12 Solutions of equation (7.12) are called *Euler's elastica*. Recent applications are to the inpainting problem (Chan *et al.*, 2002), draping of fabric (Cerda *et al.*, 2004), hot air balloon design, non-linear splines in numerical analysis (Hoschek and Lasser 1993), and computer vision (Mumford 1994).

To see the fallacy in various results appearing in the literature that we discuss below, it is vitally important to be convinced of Theorem 7.1.11.

Exercise 7.1.13 Prove Theorem 7.1.11.
Method 1 The result can be proved by direct calculation using the right hand form of the Lagrangian in equation (7.11), with the standard Euler–Lagrange operator, that is (7.2) with x as the independent variable, and then recasting the result in terms of κ and d/ds by back-substituting for u_{xxxx}, u_{xxx} and u_{xx} in terms of κ_{ss}, κ_s, κ, u_x and u (these last two factor out at the end). The calculation is lengthy but can be achieved using computer algebra software.
Method 2 Another method is to include the constraint (7.8), and to consider the Lagrangian

$$\widetilde{L} = \left[f(\kappa) - \lambda(s)(x_s^2 + u_s^2 - 1) \right] ds,$$

where $f(\kappa) = f(u_{ss}/x_s) = \kappa^2$. Calculating $0 = x_s E^x(\widetilde{L}) + u_s E^u(\widetilde{L})$ yields $0 = 2\lambda + f(\kappa) + u_s d/ds\, (f_\kappa/x_s)$ where the constant of integration has been absorbed into λ. Setting this into $E^x(\widetilde{L}) = 0$ or $E^u(\widetilde{L}) = 0$ and using both $\kappa = u_{ss}/x_s = -x_{ss}/u_s$ will, with effort, yield the result.
Method 3 Prove equation (7.18) and confirm the details of the discussion following.

Exercise 7.1.14 Show that the Euler–Lagrange equation with respect to the dependent variable u for

$$\mathcal{L}[u] = \int (\alpha + \beta\kappa^2)\, ds \tag{7.13}$$

where α and β are constants, is

$$\kappa_{ss} + \frac{1}{2}\kappa^3 - \frac{\alpha}{2\beta}\kappa = 0. \tag{7.14}$$

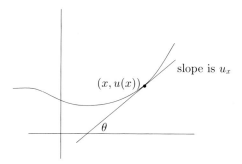

Figure 7.2 The angle $\theta = \arctan(u_x)$ satisfies $\theta_s = \kappa$, the Euclidean curvature of the curve, if s is Euclidean arc length.

The fallacious arguments involve the following line of reasoning. Consider $\theta = \arctan(u_x)$, depicted in Figure 7.2. Then if s is the Euclidean arc length,

$$\frac{\mathrm{d}}{\mathrm{d}s}\theta = \frac{1}{\sqrt{1 + u_x^2}}\frac{u_{xx}}{1 + u_x^2} = \kappa.$$

Since $\int \kappa^2 \, \mathrm{d}s = \int \theta_s^2 \, \mathrm{d}s$, and the Euler–Lagrange equation of the second 'is easily seen to be $\theta_{ss} = 0$', the extremal curves for $\int \kappa^2 \, \mathrm{d}s$ are claimed to be $\kappa = \theta_s = c$ where c is a constant. Thus, circles are taken to be the extremal curves for this problem. However, it can be seen that $\kappa = c$ is not a solution of the actual Euler–Lagrange equation, (7.12), unless $c = 0$. Further, $\theta_{ss} = 0$ is only third order, not fourth, in u. Similarly, 'spirals of Cornu', satisfying $\kappa = \alpha s + \beta$, with α and β both constant, are incorrectly derived as extremals of the variational problem,

$$\int (\theta_s^2 - \lambda\theta) \, \mathrm{d}s \qquad (7.15)$$

for a 'beam that minimises bending energy, $\int \kappa^2 \, \mathrm{d}s$, subject to having fixed amount of total angular change', so that λ is the constant Lagrange multiplier for the integral constraint, $\int \theta \, \mathrm{d}s = k$ for some constant k. In fact, the Euler–Lagrange equation for (7.15) with respect to u is

$$\kappa_{ss} + \tfrac{1}{2}\kappa^3 - \lambda\theta\kappa = 0.$$

It is readily seen that $\kappa = \theta_s = \alpha s + \beta$ is not a solution of the actual Euler–Lagrange equation unless $\alpha = \beta = 0$.

Since the method of substituting $\kappa = \theta_s$ and using s as a standard independent variable gives the wrong answer, there must be a fallacy and the question is, where? A first look at the derivation reveals one problem, that it is θ, not u, that is being varied. However, θ is a function of u_x, so varying with respect to θ and varying with respect to u should be equivalent in some sense, see Exercise 7.1.4. This accounts for the order of the purported Euler–Lagrange equation being too low. However, the real place where the above arguments fail is that the constraint $x_s^2 + u_s^2 = 1$ on the parameter s has been ignored. The constraint shows in equations (7.9) and (7.10), where one can see that there are 'hidden' occurrences of u_x in the Lagrangian $\theta_s^2 \, \mathrm{d}s$ which will add extra terms to an Euler–Lagrange equation. Incidentally, the constraint itself shows that $x = s$ cannot hold without u being identically constant; there is no way to 'pretend' that s and x are in some way equivalent. Further, equation (7.10) means one cannot pretend that s is equivalent to a free parameter t, for which $\mathrm{d}/\mathrm{d}t$ is independent of u.

7.2 Group actions on Lagrangians and Noether's First Theorem

One of the most spectacularly useful and deep results of the Calculus of Variations is Noether's Theorem. In fact, there are two theorems, one for finite dimensional Lie group actions which leads to conservation laws, and one for infinite dimensional pseudogroup actions, which leads to syzygies or differential relations, on the system of Euler–Lagrange equations. Noether proved the results in their general form (Noether (1918); modern proofs with complete formulae and a full discussion are given in Olver (1993), Section 4.4; also Theorems 5.58 and 5.66.

The simplest version of the finite dimensional result is that if a Lagrangian in one independent variable is invariant under a one parameter group action, then the Euler–Lagrange system will have a first integral which can be written down from knowledge of the infinitesimal group action alone.

Definition 7.2.1 If $g \cdot (x, u^\alpha) = (\widetilde{x}, \widetilde{u^\alpha})$ is a Lie group action, then for the variational problem $\mathcal{L}[u^\alpha] = \int_a^b L(x, u^\alpha, u_x^\alpha, \dots) \, \mathrm{d}x$, we define

$$g \cdot \mathcal{L}[u^\alpha] = \widetilde{\mathcal{L}}[\widetilde{u^\alpha}] = \int_a^b L(\widetilde{x}, \widetilde{u^\alpha}, \widetilde{u_x^\alpha}, \widetilde{u_{xx}^\alpha}, \dots) \frac{\mathrm{d}\widetilde{x}}{\mathrm{d}x} \, \mathrm{d}x,$$

where $\widetilde{u_K^\alpha}$ is the standard prolonged action on u_K^α and

$$\frac{\mathrm{d}}{\mathrm{d}x} = \frac{\partial}{\partial x} + \sum_\alpha u_x^\alpha \frac{\partial}{\partial u^\alpha} + u_{xx}^\alpha \frac{\partial}{\partial u_x^\alpha} + \cdots$$

is the total derivative operator.

It can be seen that the transformation of Lagrangians involves not only that of the function L but also that of dx; this is the reason why Lagrangians are not functions but integrands. If a Lagrangian is invariant under a one parameter group action with infinitesimals,

$$\frac{d}{d\epsilon}\Big|_{\epsilon=0}\widetilde{x} = \xi, \qquad \frac{d}{d\epsilon}\Big|_{\epsilon=0}\widetilde{u^\alpha} = \phi^\alpha, \qquad \frac{d}{d\epsilon}\Big|_{\epsilon=0}\widetilde{u_K^\alpha} = \phi^\alpha_{[K]}$$

then

$$0 = \frac{d}{d\epsilon}\Big|_{\epsilon=0} L(\widetilde{x}, \widetilde{u^\alpha}, \widetilde{u_x^\alpha}, \dots)\frac{d\widetilde{x}}{dx}$$

$$= \frac{\partial L}{\partial x}\xi + \sum_\alpha \frac{\partial L}{\partial u^\alpha}\phi + \frac{\partial L}{\partial u_x^\alpha}\phi^\alpha_{[x]} + \cdots + L\frac{d\xi}{dx}.$$

A standard argument, strongly reminiscent of the calculation of $E(L)$ but keeping track of the boundary terms performed earlier, (7.1), and using the recurrence relations derived in Exercise 1.6.17 for the infinitesimals, proves the following results.

Theorem 7.2.2 (Noether, simplest case) *If the order one Lagrangian $L(x, u^\alpha, u_x^\alpha)\,dx$ is invariant under the induced one parameter group action on (x, u^α) with infinitesimals*

$$\frac{d}{d\epsilon}\Big|_{\epsilon=0}\widetilde{x} = \xi(x, u), \qquad \frac{d}{d\epsilon}\Big|_{\epsilon=0}\widetilde{u^\alpha} = \phi^\alpha(x, u), \qquad \alpha = 1, \dots, q$$

where ϵ is the group parameter, then

$$0 = \sum_\alpha E^\alpha(L)(\phi^\alpha - u_x^\alpha\xi) + \frac{d}{dx}\left(L\xi + \sum_\alpha \frac{\partial L}{\partial u_x^\alpha}(\phi^\alpha - u_x^\alpha\xi)\right). \qquad (7.16)$$

Hence a first integral of the Euler–Lagrange system $\{E^\alpha(L) = 0\}$ is

$$L\xi + \sum_\alpha \frac{\partial L}{\partial u_x^\alpha}(\phi^\alpha - u_x^\alpha\xi) = k$$

where k is a constant.

Noether's Theorem for higher order Lagrangians in the one dimensional case is given in Exercise 7.2.8. Since these are identities for all curves $(x, u(x))$, if u is a solution of $E(L) = 0$, the expression inside the total derivative must be constant on that solution, and hence is a first integral of $E(L) = 0$. The power of even these simplest cases of Noether's Theorem can now be appreciated; knowing only the infinitesimals, the first integral can be computed symbolically with no effort at integration at all.

Exercise 7.2.3 Consider the group action

$$\tilde{x} = \exp(\epsilon)x + k, \qquad \tilde{u} = \exp(2\epsilon)u$$

where ϵ and k are the group parameters. Show the general form of a Lagrangian invariant under this action is

$$L\left(\sigma, \mathcal{D}\sigma, \mathcal{D}^2\sigma, \ldots\right) \frac{1}{\sqrt{u}}\, dx,$$

where

$$\sigma = \frac{u_x}{\sqrt{u}}, \qquad \mathcal{D} = \sqrt{u}\,\frac{d}{dx}.$$

By applying Noether's Theorem for each independent parameter, find two first integrals of $E(L) = 0$ where

$$L\, dx = \frac{u_x^2}{u^{3/2}}\, dx.$$

Hence show that solutions of $E(L) = 0$ are of the form $u(x) = (\alpha x + \beta)^4$ where α and β are constants, without using any integration. Show that the group action induces an action on the solution curves, that is, induces an action on the constants of integration.

Exercise 7.2.4 Show that

$$\tilde{x} = x, \qquad \tilde{u} = \frac{\exp(\epsilon)u}{\sinh(\epsilon)u + \exp(-\epsilon)}$$

is a one parameter group action. Find the induced action on u_x. Show that $u_x/(u(u-2))$ is an invariant, and thus write down the general form of a Lagrangian which is invariant under the action.

Exercise 7.2.5 Show that if a Lagrangian is of first order and satisfies $\partial L/\partial x = 0$, so that $L\, dx = L(u, u_x)dx$, then an integral of $E(L) = 0$ is

$$L - u_x \frac{\partial L}{\partial u_x} = k.$$

Apply the result to $\sqrt{1 + u_x^2}\, dx$. What is the result for a second order Lagrangian satisfying the same condition? Hint: the Lagrangian is invariant under the action $\tilde{x} = x + \epsilon, \tilde{u} = u$.

Exercise 7.2.6 The *brachistochrone* is a curve which is extremal for the variational problem

$$\mathcal{L}[u] = \int \left(\frac{1 + u_x^2}{u}\right)^{1/2} dx.$$

Show that the curve satisfies

$$u(1 + u_x^2) = k$$

where k is a constant. Using the substitution $u_x = \tan \psi(x)$, show that

$$x = -k(\tfrac{1}{2} \sin(2\psi) + \psi) + c$$

$$u = \tfrac{1}{2} k(1 + \cos(2\psi)),$$

where c is a constant, and hence sketch the brachistochrone as a curve in the (x, u) plane parametrised by ψ.

Exercise 7.2.7 Consider the two body (Kepler's) problem in Exercise 7.1.6. Show that the Lagrangian is invariant under four different one parameter actions, namely, translation in t, and rotations about the x-axis, the y-axis and the z-axis. Applying Noether's Theorem to each of the one parameter actions in turn, show that the Euler–Lagrange system has four first integrals,

$$\mathcal{E} = \tfrac{1}{2}(x_t^2 + y_t^2 + z_t^2) + \frac{k}{(x^2 + y^2 + z^2)^{1/2}}$$

$$\nu_x = zy_t - yz_t$$
$$\nu_y = xz_t - zx_t$$
$$\nu_z = yx_t - xy_t$$

where $\mathcal{E}$ and the ν_i are constants. The first comes from translation in time and is conservation of energy. The next three come from the rotation actions and comprise conservation of angular momentum. Show that an extremal curve $t \mapsto (x(t), y(t), z(t))$ lies on a plane through the origin normal to (ν_x, ν_y, ν_z). Show that the group action sends solutions of the Euler–Lagrange system to solutions, and induces an action on the constant $\mathcal{E}$ and on the vector (ν_x, ν_y, ν_z) (in fact $\mathcal{E}$ is invariant). By rotating the coordinates, we can assume the solution lies on the plane $z \equiv 0$, that is, $(\nu_x, \nu_y, \nu_z) = (0, 0, \nu)$. Write the reduced system of equations, that is with $z \equiv 0$, in terms of $\rho = x^2 + y^2$ and $\theta = \arctan(y/x)$ and the constants k, $\mathcal{E}$ and ν. Hint: calculate ρ_{tt} and use conservation of energy to simplify. Note: the fact we obtain an autonomous ODE for the rotation invariant ρ means that the 'reduced' Euler–Lagrange system still retains two symmetries; one rotational symmetry as well as translation in time. The other two rotational symmetries have been 'normalised' by selecting the plane containing the solution.

Exercise 7.2.8 (Noether's Theorem, higher order Lagrangians) Suppose that the one dimensional but arbitrary order Lagrangian

$$L(x, u, u_x, u_{xx}, \dots) \, dx$$

is invariant under a one parameter group action induced by $\epsilon \cdot (x, u) = (\widetilde{x}, \widetilde{u})$ with infinitesimals

$$\frac{\mathrm{d}}{\mathrm{d}\epsilon}\bigg|_{\epsilon=0} \widetilde{x} = \xi(x, u), \qquad \frac{\mathrm{d}}{\mathrm{d}\epsilon}\bigg|_{\epsilon=0} \widetilde{u} = \phi(x, u).$$

Show that

$$0 = QE(L) + \frac{\mathrm{d}}{\mathrm{d}x}\left(L\xi + \sum_{m=1}^{m-1}\sum_{k=0}^{m-1}(-1)^k \left(\frac{\mathrm{d}^k}{\mathrm{d}x^k}\frac{\partial L}{\partial u_m}\right)\left(\frac{\mathrm{d}^{m-1-k}Q}{\mathrm{d}x^{m-1-k}}\right)\right)$$

where $Q = \phi - u_x\xi$ and

$$u_m = \frac{\mathrm{d}^m}{\mathrm{d}x^m}u.$$

Hence deduce the first integral of $E(L) = 0$ given by the symmetry. By considering the group action to be translation in x, with infinitesimals $\phi = 0$ and $\xi = 1$, show that $u_x E^u(L)$ is a total derivative when $\partial L/\partial x = 0$, specifically,

$$u_x E(L) = \frac{\mathrm{d}}{\mathrm{d}x}\left(L - \sum_{m=1}^{m-1}\sum_{k=0}^{m-1}(-1)^k \left(\frac{\mathrm{d}^k}{\mathrm{d}x^k}\frac{\partial L}{\partial u_m}\right)u_{m-k}\right). \qquad (7.17)$$

Exercise 7.2.9 Use Exercise 7.2.8 to extend the result of Theorem 7.2.2 to nth order Lagrangians with more than one dependent variable.

7.2.1 Moving frames and Noether's Theorem, the appetizer

In this section, we explore Noether's Theorem for the most famous Lagrangian involving Euclidean curvature and arc length. We find, amazingly, that the formulae for Noether's first integrals calculate a representation of a frame.

Using the right hand form of the Lagrangian in equation (7.11), the three first integrals of equation (7.12), obtained by evaluating the integral obtained in Exercise 7.2.8 for the one parameter group actions corresponding to each independent group parameter in turn, can be written in the form

$$\begin{pmatrix} c_1 \\ c_2 \\ c_3 \end{pmatrix} = \begin{pmatrix} \dfrac{1}{\sqrt{1+u_x^2}} & -\dfrac{u_x}{\sqrt{1+u_x^2}} & 0 \\[2mm] \dfrac{u_x}{\sqrt{1+u_x^2}} & \dfrac{1}{\sqrt{1+u_x^2}} & 0 \\[2mm] \dfrac{xu_x - u}{\sqrt{1+u_x^2}} & \dfrac{uu_x + x}{\sqrt{1+u_x^2}} & 1 \end{pmatrix} \begin{pmatrix} -\kappa^2 \\ -2\kappa_s \\ 2\kappa \end{pmatrix} \qquad (7.18)$$

where the c_i are the constants of integration. The first component comes from translation in x, the second from translation in u and the third from rotation in the (x, u) plane about the origin.

Let us denote by $B = B(x, u(x))$ the 3×3 matrix appearing in equation
(7.18). Remarkably, B is equivariant with respect to the Euclidean action.
Further, using the invariant differential operator, d/ds, the matrix $B^{-1}B_s$ is
invariant, indeed, we have

$$B^{-1}B_s = \begin{pmatrix} 0 & -\kappa & 0 \\ \kappa & 0 & 0 \\ 0 & 1 & 0 \end{pmatrix}. \tag{7.19}$$

Differentiating equation (7.18) with respect to s, and using (7.19) to simplify,
yields another proof of Theorem 7.1.11, that is, extremal curves for (7.11)
satisfy equation (7.12).

Exercise 7.2.10 Show that the 3×3 matrix, denoted as $B(x, u)$, appearing in
equation (7.18) is equivariant with respect to the standard action of $SO(2) \ltimes \mathbb{R}^2$
on curves on the plane. Indeed, show that

$$B(x\cos\theta - u\sin\theta + a, x\sin\theta + u\cos\theta + b)$$

$$= \begin{pmatrix} \cos\theta & -\sin\theta & 0 \\ \sin\theta & \cos\theta & 0 \\ a\sin\theta - b\cos\theta & b\sin\theta + a\cos\theta & 1 \end{pmatrix} B(x, u)$$

$$= \mathcal{R}(\theta, a, b)B(x, u),$$

where the last equation defines $\mathcal{R}(\theta, a, b)$. Show that $(\theta, a, b) \mapsto \mathcal{R}(\theta, a, b)$ is
a matrix representation of $SO(2) \ltimes \mathbb{R}^2$. Compare it to the Adjoint representa-
tion obtained in Exercise 3.3.9. Show that if ρ is the frame for the standard
action of $SO(2) \ltimes \mathbb{R}^2$ on curves in the plane, determined by the normalisation
equations $\widetilde{x} = \widetilde{u} = \widetilde{u}_x = 0$, then $B = \mathcal{R}(\rho)^{-1}$. Since in this case the $\mathcal{A}d$ representation is faithful, the group action can be used to simplify the integration
problem.

There are several important consequences of the equivariance of the matrix
B. The first is that an explicit induced group action on the first integrals is
obtained. This result was known in infinitesimal form, see for example, Olver
(1993), page 341. The second important remark is that B is, by definition, a
frame for the Euclidean action on curves in the plane, since it is an equivariant
map from the space on which the group acts to (a matrix representation of)
the group. Thus the formula giving the first integrals via Noether's Theorem
has calculated a frame without any input of moving frame theory. Indeed,
the invariants in equation (7.18) are historical while the formulae for the first
integrals are in the original coordinates; they are blind to any knowledge of the
invariants.

If we regard κ as known once we have solved the invariantised Euler–Lagrange equation, the three components of equation (7.18) together with $x_s^2 + u_s^2 = 1$ can be viewed as four equations for four unknowns, x, u, x_s and u_s. In this particular case, it is possible to solve for u in terms of x, the c_i and κ with a single integration, see Exercise 7.2.11.

Exercise 7.2.11 Consider the three components of equation (7.18) as three equations for x, u and u_x in terms of the c_i and κ. Show that

$$c_1^2 + c_2^2 = \kappa^4 + 4\kappa_s^2.$$

Relate this equation to a first integral of the Euler–Lagrange equation, $\kappa_{ss} + \frac{1}{2}\kappa^3 = 0$, and deduce there is an equation for the c_i in terms of initial data chosen for the Euler–Lagrange equation. Show further that u can be obtained with a single integration by finding an algebraic consequence of the equations that does not contain x_s or u_s. Hint: set $x_s = 1/\sqrt{1 + u_x^2}$ and $u_s = u_x/\sqrt{1 + u_x^2}$ into the components of equation (7.18) together with $x_s^2 + u_s^2 = 1$ and eliminate u_s, x_s.

Equation (7.18) and the results of Exercise 7.2.10 are not flukes, but rather an example of a general result, see Theorem 7.4.1.

7.3 Calculating invariantised Euler–Lagrange equations directly

In the following, we find the Euler–Lagrange equation of an invariant Lagrangian such as (7.11) directly in terms of the invariants. This problem was considered by Kogan and Olver (2003) using a variational tricomplex. Here we show how the result can be achieved using the invariant calculus developed in Chapters 4 and 5. We use computations that are the direct analogues of those in the standard case and which explicitly include constraints such as the parameter being arc length. While there will be a computational advantage in high order cases to use the invariant calculus, the main advantage is in understanding the form the Euler–Lagrange equation takes; the syzygies between the invariants plays a significant role, as we shall see.

We saw earlier that given $\mathcal{L}[u] = \int L \, dx$, the two steps that calculate $E(L) = 0$ are

(i) calculate

$$\frac{d}{d\epsilon}\Big|_{\epsilon=0} \mathcal{L}[u + \epsilon v]$$

(ii) integrate by parts.

In the simplest case of unconstrained invariant independent variables, there are three steps to calculate $E(L) = 0$ directly in terms of the invariants:

(i) calculate an invariantised analogue of

$$\frac{\mathrm{d}}{\mathrm{d}\epsilon}\Big|_{\epsilon=0} \mathcal{L}[u + \epsilon v]$$

(ii) apply a syzygy
(iii) integrate by parts.

If the independent variables are constrained, for example the curve parameter is Euclidean arc length, or are not invariant, there are an additional two preprocessing steps,

(i) reparametrise so that the independent variables are both invariant and unconstrained, this means that the former independent variables are now dependent variables,
(ii) include any constraints into the Lagrangian with a Lagrange multiplier. It is also possible to include companion equations, which effectively fix the parametrisation, as constraints; chosen with care, these can dramatically simplify the calculations.

The companion equations need to be compatible in the sense that the system comprising the resulting Euler–Lagrange system and the companion equations have the same solution set as the original Euler–Lagrange system. Just as when we used this technique in Chapter 6, companion equations can be used symbolically and need never actually be solved, nor do they need to be realised in the original variables; only their invariantised forms are needed.

Remark 7.3.1 Although the computations that follow can be adapted to non-commuting invariant operators, reparametrisation ensures the resulting Euler–Lagrange equations are differential equations employing standard derivative operators.

In what follows, we will need to use the adjoint of the differential operators defining the syzygies.

Definition 7.3.2 If $\mathcal{A}$ is the differential operator

$$\mathcal{A}(f) = \sum_K A_K \frac{\mathrm{d}^{|K|}}{\mathrm{d}x^K},$$

where A_K are functions of the independent variables and K is the (multi)index of differentiation, we define the *adjoint* of $\mathcal{A}$, denoted by $\mathcal{A}^*$, to be

$$\mathcal{A}^*(f) = \sum_K (-1)^{|K|} \frac{\mathrm{d}^{|K|}}{\mathrm{d}x^K} (A_K \, f).$$

The adjoint operator satisfies

$$\int f \mathcal{A}(g) = \int \mathcal{A}^*(f) g + \text{B.T.s}$$

where 'B.T.s' stands for boundary terms.

Exercise 7.3.3 Let $P[u] = P(x, u, u_x, \ldots)$ and define D_P to be the operator such that

$$\mathrm{D}_P(Q) = \left. \frac{\mathrm{d}}{\mathrm{d}\epsilon} \right|_{\epsilon=0} P[u + \epsilon Q].$$

Show that $E(P) = \mathrm{D}_P^*(1)$.

7.3.1 The case of invariant, unconstrained independent variables

To begin with, assume we have a single invariant, unconstrained independent variable x and a single dependent variable, u.

To obtain the invariantised analogue of

$$\left. \frac{\mathrm{d}}{\mathrm{d}\epsilon} \right|_{\epsilon=0} \mathcal{L}[u + \epsilon v],$$

we introduce τ, a dummy invariant independent variable, and take a clue from the observation that

$$\left. \frac{\mathrm{d}}{\mathrm{d}\epsilon} \right|_{\epsilon=0} \mathcal{L}[u + \epsilon v] = \left. \frac{\mathrm{d}}{\mathrm{d}\tau} \right|_{u_\tau = v} \mathcal{L}[u].$$

Since both $\tilde{\tau} = \tau$, and $\tilde{x} = x$, by construction and hypothesis, we have

$$\mathcal{D}_\tau = \frac{\mathrm{d}}{\mathrm{d}\tau}, \quad \mathcal{D}_x = \frac{\mathrm{d}}{\mathrm{d}x}, \quad \left[\frac{\mathrm{d}}{\mathrm{d}\tau}, \frac{\mathrm{d}}{\mathrm{d}x} \right] = 0. \tag{7.20}$$

We take the moving frame with respect to the x derivatives of the dependent variable only, and assume the frame is chosen so that there are only two generating invariants, one 'in the τ direction', $I_2 = \iota(u_\tau)$ and one in the x direction (this requirement will be relaxed later),

$$\sigma = \iota(u_K)$$

where K is an index of differentiation in the independent variable x only. The syzygy between σ and I_2 can be written as

$$\mathcal{D}_\tau \sigma = \mathcal{H} I_2 \tag{7.21}$$

where $\mathcal{H}$ is an operator with coefficients that are functions of σ and its derivatives with respect to x.

The outline of the process of finding the Euler–Lagrange equations directly in terms of the invariants is as follows. Since τ is an introduced dummy variable, the invariant Lagrangian will be a function of σ and its derivatives with respect to x. Hence we have

$$\frac{d}{d\tau} \int L(x, \sigma, \mathcal{D}_x\sigma, \mathcal{D}_x^2\sigma, \dots) \, dx$$

$$= \int \left(\frac{\partial L}{\partial \sigma} + \frac{\partial L}{\partial \mathcal{D}_x\sigma} \mathcal{D}_x + \cdots \right) \mathcal{D}_\tau\sigma \, dx$$

$$= \int \left(\frac{\partial L}{\partial \sigma} - \mathcal{D}_x \frac{\partial L}{\partial \mathcal{D}_x\sigma} + \mathcal{D}_x^2 \frac{\partial L}{\partial \mathcal{D}_x^2\sigma} + \cdots \right) \mathcal{H}I_2 \, dx + \text{ B.T.s}$$

$$= \int \mathcal{H}^* \left(\frac{\partial L}{\partial \sigma} - \mathcal{D}_x \frac{\partial L}{\partial \mathcal{D}_x\sigma} + \mathcal{D}_x^2 \frac{\partial L}{\partial \mathcal{D}_x^2\sigma} + \cdots \right) I_2 \, dx + \text{ more B.T.s}$$

$$= \int \mathcal{H}^* E^\sigma(L) I_2 \, dx + \text{ the B.T.s}$$

where 'B.T.s' stands for boundary terms and $E^\sigma(L)$ is the Euler–Lagrange operator applied to L treated as a function of σ and its x derivatives. To do the calculation, we have used equation (7.20) in the first line, performed a first set of integration by parts and then used equation (7.21) in the second, and finally a second integration by parts in the third line, so that $\mathcal{H}^*$ is the standard adjoint of the operator $\mathcal{H}$. We now note that I_2 has the factor u_τ (this follows from the Replacement Theorem) which is the independent variation in the dependent variable. It follows from the Fundamental Lemma of the Calculus of Variations that the coefficient of I_2^u must be zero,

$$\mathcal{H}^* \left(\frac{\partial L}{\partial \sigma} - \mathcal{D}_x \frac{\partial L}{\partial \mathcal{D}_x\sigma} + \mathcal{D}_x^2 \frac{\partial L}{\partial \mathcal{D}_x^2\sigma} + \cdots \right) = 0 \qquad (7.22)$$

and thus this is the sought for invariantised Euler–Lagrange equation. We have thus proved the following theorem.

Theorem 7.3.4 *If $\mathcal{L}[u] = \int L(\sigma, \sigma_x, \dots) \, dx$ is a variational problem for planar curves $(x, u(x))$ where σ is the generating invariant of a group action on curves such that $\widetilde{x} = x$, then the Euler–Lagrange equation is*

$$\mathcal{H}^* E^\sigma(L) = 0 \qquad (7.23)$$

where $\mathcal{H}$ is the operator in the syzygy (7.21).

Example 7.3.5 For our first set of examples, we take our old friend the $SL(2)$ action,

$$\widetilde{x} = x, \qquad \widetilde{u} = \frac{au + b}{cu + d}, \qquad ad - bc = 1$$

with the normalisation equations

$$\tilde{u} = 0, \qquad \tilde{u}_x = 1, \qquad \tilde{u}_{xx} = 0.$$

The generating invariant is I_{111}^u which we know is the Schwarzian derivative, $\{u; x\}$ and thus we set

$$\sigma = \frac{u_{xxx}}{u_x} - \frac{3}{2}\frac{u_{xx}^2}{u_x^2}.$$

As in the derivation above, we introduce the dummy independent variable τ, to obtain a new invariant $I_2 = \iota(u_\tau) = u_\tau/u_x$, with the syzygy,

$$\mathcal{D}_\tau \sigma = \mathcal{H}I_2 = \left(\mathcal{D}_x^3 + 2\sigma\mathcal{D}_x + \sigma_x\right)I_2. \tag{7.24}$$

The adjoint of $\mathcal{H}$ is

$$\mathcal{H}^* = -\mathcal{D}_x^3 - 2\sigma\mathcal{D}_x - \sigma_x = -\mathcal{H}.$$

Following the calculation above for $L(\sigma, \sigma_x, \dots)\,dx$ the Euler–Lagrange equation is $\mathcal{H}^*E^\sigma(L) = 0$. Thus, for example, for $L\,dx = \sigma^2\,dx$, we obtain the Euler–Lagrange equation

$$0 = \mathcal{H}^*(\sigma) = -\sigma_{xxx} - 2\sigma\sigma_x - \sigma_x\sigma = -\sigma_{xxx} - 3\sigma\sigma_x$$

which can also be verified directly. Note that since σ is a third order invariant, we expect the Euler–Lagrange equation to be of order 6, that is in terms of σ_{xxx}, and this we obtain. Further, we see that the terms in the Euler–Lagrange equation that are additional to $E^\sigma(L)$ come from the syzygy (7.24), as promised earlier.

Exercise 7.3.6 Find the Euler–Lagrange equations for the Lagrangians $\sigma\,dx$, $\sigma_x^2\,dx$ and $x\sigma^2\,dx$. Comment on the order of the Euler–Lagrange equation for $\sigma\,dx$. Hint: see Exercise 7.1.3.

It is not hard to see how to generalise the calculation to more than one dependent variable. The new twist is that the syzygy (7.24) becomes a matrix equation. Since τ is a dummy variable used to effect the variation, the frame is obtained with respect to the x derivatives of the dependent variables only. Thus, if the generating x-derivative invariant of u^α is denoted σ_α and $\iota(u_\tau^\alpha) = I_2^\alpha$, for $\alpha = 1, \dots, q$, then the syzygy between the σ_α and the I_2^α can be written as

$$\mathcal{D}_\tau \begin{pmatrix} \sigma_1 \\ \vdots \\ \sigma_q \end{pmatrix} = \begin{pmatrix} \mathcal{H}_{11} & \mathcal{H}_{12} & \cdots & \mathcal{H}_{1q} \\ \vdots & \vdots & \ddots & \vdots \\ \mathcal{H}_{q1} & \mathcal{H}_{q2} & \cdots & \mathcal{H}_{qq} \end{pmatrix} \begin{pmatrix} I_2^1 \\ \vdots \\ I_2^q \end{pmatrix}. \tag{7.25}$$

Example 7.3.7 Suppose we take the product $SL(2)$ action on sets of pairs of curves in the plane, $(x, r(x), s(x))$,

$$\widetilde{x} = x, \qquad \widetilde{r} = \frac{ar + b}{cr + d}, \qquad \widetilde{s} = \frac{as + b}{cs + d}, \qquad ad - bc = 1$$

with the normalisation equations, $\widetilde{r} = 0$, $\widetilde{r}_x = 1$, $\widetilde{s} = 1$, so that the generators with respect to the x derivatives are

$$\chi = \iota(r_{xx}), \qquad \eta = \iota(s_x).$$

Set $I_2^r = \iota(r_\tau)$ and $I_2^s = \iota(s_\tau)$. Then the syzygies can be put in the form

$$\mathcal{D}_\tau \begin{pmatrix} \chi \\ \eta \end{pmatrix} = \begin{pmatrix} \mathcal{D}_x^2 + 2\eta + \chi_x + (2 + \chi)\mathcal{D}_x & -2 \\ \eta\mathcal{D}_x + 2\eta + \eta\chi & \mathcal{D}_x - \chi - 2 \end{pmatrix} \begin{pmatrix} I_2^r \\ I_2^s \end{pmatrix}$$

where $\mathcal{D}_x = \mathrm{d}/\mathrm{d}x$. Following the line of calculation above, we obtain for the Lagrangian, $L\,\mathrm{d}x = L(x, \chi, \eta, \chi_x, \eta_x, \dots)\,\mathrm{d}x$,

$$\frac{\mathrm{d}}{\mathrm{d}\tau} \int L\,\mathrm{d}x = \int E^\chi(L)\mathcal{D}_\tau\chi + E^\eta(L)\mathcal{D}_\tau\eta\,\mathrm{d}x + \text{B.T.s}$$

$$= \int \left[(\mathcal{D}_x^2 + 2\eta + \chi_x + (2 + \chi)\mathcal{D}_x)^* E^\chi(L) \right.$$

$$+ (\eta\mathcal{D}_x + 2\eta + \eta\chi)^* E^\eta(L) \Big] I_2^r$$

$$+ \left[-2E^\chi(L) + (\mathcal{D}_x - \chi - 2)^* E^\eta(L) \right] I_2^s\,\mathrm{d}x + \text{B.T.s}$$

so that the Euler–Lagrange system is, in matrix form,

$$0 = \begin{pmatrix} \mathcal{D}_x^2 + 2\eta - (2 + \chi)\mathcal{D}_x & -\eta\mathcal{D}_x - \eta_x + 2\eta + \eta\chi \\ -2 & -\mathcal{D}_x - \chi - 2 \end{pmatrix} \begin{pmatrix} E^\chi(L) \\ E^\eta(L) \end{pmatrix}.$$

To generalise the calculation of the invariantised Euler–Lagrange system to the general case of p invariant, unconstrained independent variables and q dependent variables, one uses the syzygies between all the generating invariants $\sigma_i = \iota(u_{K_i}^{\alpha_i})$ for $i = 1, \dots, N$ and the $\iota(u_\tau^\alpha)$, $\alpha = 1, \dots, q$, as above, but now the new twist is that there are additional generating syzygies between the σ_i. These need to be included as constraints, each with its own Lagrange multiplier function, in the Lagrangian: this last generalisation includes the case of more than one generating invariant per dependent variable. We leave this straightforward generalisation to the reader to explore.

7.3.2 The case of non-invariant independent variables

We next consider the case of non-invariant independent variables. Suppose we have $x = (x_1, \dots, x_p)$, and $\widetilde{x}_i \neq x_i$ for at least one i. Introduce $t = (t_1, \dots, t_p)$

and reparametrise the surface described by $(x_1, \ldots, x_p, u^1(x), \ldots, u^q(x))$ to be given instead by

$$(t_1, \ldots, t_p, x_1(t), \ldots, x_p(t), u^1(t), \ldots, u^q(t)).$$

Actually, only as many new parameters are needed as there are non-invariant x_i. We assume that $\widetilde{t}_i = t_i$ for all i in order to obtain invariant independent variables. The main benefit is that the resulting differential equations will have derivative operators in the $\partial/\partial t_i$ and thus can be treated by standard methods. Thus the original independent variables become dependent variables. The twist is that we may introduce 'companion equations' for the x_i; these may be thought of as fixing the parametrisation in some sense. They are chosen to simplify the calculations but care must be taken to ensure no loss of solutions results. The companion equations are then taken to be *constraints* on the reparametrised Lagrangian. Just as when we integrated invariant ordinary differential equations using this exact same mechanism in Chapter 6, we will *not* need to solve the companion equations in order to solve the resulting Euler–Lagrange system, nor will we need to know even what they look like in the original variables; it is enough to show that the resulting solution set has the same dimension solution space as the original Euler–Lagrange system. If companion equations are not used, the Lagrangian will be invariant under a pseudogroup of coordinate changes and then by Noether's Second Theorem, the Euler–Lagrange equations will not be independent but will satisfy a differential relation.

Example 7.3.8 We consider another old friend, the $SL(2)$ action under which the Chazy equation, equation (6.16), is invariant. Recall this action on $(x, u(x))$ is given by

$$\widetilde{x} = \frac{ax+b}{cx+d}, \qquad \widetilde{u} = 6c(cx+d) + (cx+d)^2 u, \qquad ad - bc = 1. \quad (7.26)$$

Set $x = x(t)$, $u = u(t)$, and introduce the dummy variable τ so that

$$\mathcal{D}_\tau = \frac{\mathrm{d}}{\mathrm{d}\tau}, \qquad \mathcal{D}_t = \frac{\mathrm{d}}{\mathrm{d}x}, \qquad \left[\frac{\mathrm{d}}{\mathrm{d}\tau}, \frac{\mathrm{d}}{\mathrm{d}t}\right] = 0. \quad (7.27)$$

Taking the normalisation equations to be

$$\widetilde{x} = 0, \qquad \widetilde{u} = 0, \qquad \widetilde{u}_t = 1$$

we obtain generating t-derivative invariants,

$$\kappa = I_{11}^u = \iota(u_{tt}), \qquad \eta = I_1^x = \iota(x_t)$$

τ-derivative invariants,

$$I_2^x = \iota(x_\tau), \qquad I_2^u = \iota(u_\tau)$$

and syzygies

$$\mathcal{D}_\tau \begin{pmatrix} \kappa \\ \eta \end{pmatrix} = \begin{pmatrix} \mathcal{D}_t^2 - \frac{1}{3}\eta + \kappa\mathcal{D}_t + \kappa_t & \frac{1}{3} \\ \eta\kappa + \eta\mathcal{D}_t & \mathcal{D}_t - \kappa \end{pmatrix} \begin{pmatrix} I_2^u \\ I_2^x \end{pmatrix}.$$

Suppose we have a Lagrangian that is invariant under the action in equation (7.26). Taking $\eta = 1$ for the companion equation (a choice that needs to be vindicated in the example at hand; particular Lagrangians may need a different choice), we reparametrise and apply Theorem 4.4.9, the Replacement Theorem, and obtain a variational problem of the form

$$\mathcal{L}[\kappa, \eta] = \int [L(\kappa, \kappa_t, \dots) - \lambda(t)(\eta - 1)] \, dt$$

where we have already used the equation $\eta = 1$ in the arguments of L, and where $\lambda(t)$ is the Lagrange multiplier function. We now follow the calculation above to obtain

$$\frac{\partial}{\partial\tau}\mathcal{L}[\kappa, \eta] = \frac{\partial}{\partial\tau} \int [L(\kappa, \kappa_t, \dots) - \lambda(t)(\eta - 1)] \, dt$$

$$= \int \left[E^\kappa(L)\mathcal{D}_\tau\kappa - \lambda\mathcal{D}_\tau\eta \right] dt + \text{B.T.s}$$

$$= \int \left[E^\kappa(L)\left(\mathcal{D}_2^2 - \frac{1}{3} + \kappa\mathcal{D}_t - \kappa_t\right) I_2^u + \frac{1}{3}I_2^x \right)$$

$$\qquad - \lambda(t)\left[(\kappa + \mathcal{D}_t) I_2^u + (\mathcal{D}_t - \kappa) I_2^x \right] dt + \text{B.T.s}$$

$$= \int \left[\left(\mathcal{D}_t^2 - \frac{1}{3} - \kappa\mathcal{D}_t \right) E^\kappa(L) + (\lambda_t - \kappa\lambda) \right] I_2^u$$

$$\qquad + \left[\frac{1}{3}E^\kappa(L) + \kappa\lambda + \lambda_t \right] I_2^x \, dt + \text{more B.T.s}$$

where we have already set $\eta = 1$ in the syzygy operator; it makes no difference to the final result to do so at this stage. Since I_2^u and I_2^x are the independent variations, the Euler–Lagrange equations are the coefficients of I_2^u and I_2^x in the final integrand. The two equations can be solved for λ and λ_t and the equation for λ_t can be written as

$$\mathcal{D}_t^2 E^\kappa(L) - \mathcal{D}_t\left(\kappa E^\kappa(L)\right) + \kappa_t E^\kappa(L) + 2\lambda_t = 0.$$

We now note that $\partial L/\partial t = 0$ since when a Lagrangian is reparametrised the new independent variable never appears explicitly. Thus by the final result of Exercise 7.2.8, equation (7.17), the term $\kappa_t E^\kappa(L)$ is a total derivative. Thus we can integrate this last equation in order to eliminate the derivative on λ; the constant of integration is absorbed into λ as an artefact, see Remark 7.1.9. In this way, two equations for λ are obtained, so λ can be eliminated, and we obtain the invariantised Euler–Lagrange equation to be

$$\left(\mathcal{D}_t^2 - \kappa^2 - \tfrac{2}{3}\right) E^\kappa(L) - \kappa L + \kappa \sum_{m=1}^{m-1}\sum_{k=0}^{m-1}(-1)^k \left(\frac{\mathrm{d}^k}{\mathrm{d}t^k}\frac{\partial L}{\partial \kappa_m}\right)\kappa_{m-k} = 0$$

where

$$\kappa_m = \frac{\mathrm{d}^m}{\mathrm{d}t^m}\kappa.$$

Once $\kappa(t)$ is known, the methods of Section 6.4.2 or Section 7.4 can be used to obtain the extremals in terms of $(x(t), u(t))$.

7.3.3 The case of constrained independent variables such as arc length

It can happen that the independent variable is invariant but nevertheless is not a free parameter. This is the case for Euclidean arc length. These kinds of problems can also treated by reparametrisation, and the inclusion of the appropriate constraint in the Lagrangian.

The prototypical examples are Lagrangians for curves $(x, u(x))$ in the plane which are invariant under the Euclidean group $SE(2)$. We have that $\tilde{s} = s$ but that s is constrained, in the sense that s satisfies $x_s^2 + u_s^2 = 1$. Variational problems with this Euclidean symmetry take the form,

$$\mathcal{L}[x, u] = \int L(\kappa, \kappa_s, \dots)\,\mathrm{d}s.$$

Remark 7.3.9 We do not consider Lagrangians depending explicitly on the parameter s. Any Lagrangian given initially in the $(x, u, u_x, \dots)$ coordinates and rewritten in terms of κ and $\mathrm{d}s$ will not depend explicitly on s. We note arc length is an integral, not a differential, invariant.

If we take the normalisation equations

$$\tilde{x} = 0, \qquad \tilde{u} = 0, \qquad \tilde{u}_s = 0$$

the generating invariants are

$$\sigma = I_{11}^u = \iota(u_{ss}), \qquad \eta = I_1^x = \iota(x_s).$$

The replacement rule gives $\kappa = (I_{11}^u I_1^x - I_1^x I_1^u)/(I_1^x)^3$, and the arc length constraint becomes $I_1^x = 1$. Since here, $\mathcal{D}_s I_1^x = I_{11}^x$, the curvature

$$\kappa = I_{11}^u = \sigma \quad \text{when } I_1^x \equiv 1.$$

Introducing the dummy independent variable τ we obtain two new generating invariants,

$$I_2^u = \iota(u_\tau), \qquad I_2^x = \iota(x_\tau)$$

and syzygies

$$\mathcal{D}_\tau \begin{pmatrix} \kappa \\ \eta \end{pmatrix} = \mathcal{H} \begin{pmatrix} I_2^u \\ I_2^x \end{pmatrix} \tag{7.28}$$

where

$$\mathcal{H} = \begin{pmatrix} \mathcal{D}_s^2 - \dfrac{\eta_s}{\eta}\mathcal{D}_s - \left(\dfrac{\kappa}{\eta}\right)^2 & -2\dfrac{\kappa\eta_s}{\eta^2} + \dfrac{\kappa_s}{\eta} + 2\dfrac{\kappa}{\eta}\mathcal{D}_s \\ -\dfrac{\kappa}{\eta} & \mathcal{D}_s \end{pmatrix}. \tag{7.29}$$

We consider the constrained Lagrangian,

$$\mathcal{L} = \int \left[L(\kappa, \kappa_s, \dots) - \lambda(s)(\eta - 1) \right] \mathrm{d}s.$$

Setting

$$\kappa_m = \frac{\mathrm{d}^m}{\mathrm{d}s^m}\kappa,$$

the calculation of the invariant Euler–Lagrange system is as follows:

$$\mathcal{D}_\tau \mathcal{L} = \int \sum_m \frac{\partial L}{\partial \kappa_m} \frac{\mathrm{d}^m}{\mathrm{d}s^m} \mathcal{D}_\tau \kappa - \lambda(s)\mathcal{D}_\tau \eta$$

$$= \int E^\kappa(L)\mathcal{D}_\tau \kappa - \lambda(s)\mathcal{D}_\tau \eta + \text{B.T.s}$$

$$= \int E^\kappa(L)\left[\left(\mathcal{D}_s^2 - \kappa^2\right)I_2^u + (\kappa_s + 2\kappa\mathcal{D}_s)I_2^x \right]$$

$$\qquad - \lambda(s)\left[-\kappa I_2^u + \mathcal{D}_2 I_2^x \right] + \text{B.T.s}$$

$$= \int \left[(\mathcal{D}_s^2 - \kappa^2)E^\kappa(L) + \kappa\lambda \right] I_2^u$$

$$\qquad + \left[\kappa_s E^\kappa(L) - 2(\kappa E^\kappa(L))_s + \lambda_s \right] I_2^x + \text{ more B.T.s},$$

where we have already set $\eta \equiv 1$ in the operator $\mathcal{H}$; it makes no difference to the end result to do so at this stage. Thus the Euler–Lagrange system is, in addition to $\eta \equiv 1$,

$$0 = (\mathcal{D}_s^2 - \kappa^2)E^\kappa(L) + \kappa\lambda,$$

$$0 = \kappa_s E^\kappa(L) - 2\left[\kappa E^\kappa(L) \right]_s + \lambda_s. \tag{7.30}$$

Since L does not depend explicitly on s, by equation (7.17) in Exercise 7.2.8, $\kappa_s E^\kappa(L)$ is a total derivative (setting κ for u and s for x; the proof is unaffected by the constraint on s). Hence we can integrate the second equation, absorb the constant of integration into λ as in the non-invariantised case, and then eliminate λ; see Exercise 7.1.7 and Remark 7.1.9. In this way we obtain the differential equation for κ that is equivalent to the Euler–Lagrange equation with respect to u. The result after collecting terms is

$$\left(\mathcal{D}_s^2 + \kappa^2\right) E^\kappa(L) + \kappa \left[-L + \sum_{m=1}^{m-1} \sum_{k=0}^{} (-1)^k \left(\frac{d^k}{ds^k} \frac{\partial L}{\partial \kappa_m} \right) \kappa_{m-k} \right] = 0. \quad (7.31)$$

Once $\kappa(s)$ is known, the methods of Chapter 6 can be used to obtain the extremal curves in (x, u) space.

Exercise 7.3.10 Verify the following using both the formula (7.31) and directly; the use of computer algebra is strongly recommended for the latter.

Lagrangian	Euler–Lagrange equation
$\kappa^2 \, ds$	$2\kappa_{ss} + \kappa^3 = 0$
$\kappa_s^2 \, ds$	$-2\kappa_{ssss} - 2\kappa^2 \kappa_{ss} + \kappa \kappa_s^2 = 0$
$\kappa_{ss}^2 \, ds$	$2\kappa_6 + 2\kappa_4 \kappa^2 - 2\kappa \kappa_s \kappa_3 + \kappa \kappa_{ss}^2 = 0.$

where κ_m is the mth derivative of κ with respect to s.

Exercise 7.3.11 If $L \, ds = \kappa_s \kappa^2 \, ds$, show both $E^\kappa(L)$ and $E^u(L)$ are identically zero. Explain.

7.3.4 The 'mumbo jumbo'-free rigid body

I have always been struck by the high level of jargon and the strange, awkward and mystifying definitions, constructions and calculations that surround the mathematical treatment of a rigid body. Here we show how the three steps used above, namely, calculate the variation, apply a syzygy, and integrate by parts, make calculating the Euler–Lagrange equations for a rigid body both straightforward and painless.

It is always important, when studying a physical problem, to divide the physics from the mathematics. The assumptions about the model, and the derivation of the Lagrangian for the model, are physics. The rest is mathematics. Suppose the rigid body is composed of particles of mass m_α located at $\mathbf{x}_\alpha(t)$. The physics of any body made up of such particles, not subject to any forces, is that the Lagrangian is given by

$$L \, dt = \sum \tfrac{1}{2} m_\alpha \langle \dot{\mathbf{x}}_\alpha, \dot{\mathbf{x}}_\alpha \rangle, \, dt \quad (7.32)$$

where $\dot{\mathbf{x}} = d\mathbf{x}/dt$ and $\langle \mathbf{x}, \mathbf{x} \rangle$ is the standard inner product. Implicit in this Lagrangian is that the equations of motion will be invariant under translation in time and space, because those actions leave the Lagrangian invariant. The calculations that follow are greatly simplified if the origin of space time is set to be such that

$$\sum m_\alpha \mathbf{x}_\alpha(0) = 0. \tag{7.33}$$

This can be achieved by the constant translation,

$$\mathbf{x}_\alpha(t) \mapsto \mathbf{x}_\alpha(t) - \frac{1}{\sum m_\alpha} \sum m_\alpha \mathbf{x}_\alpha(0).$$

Remark 7.3.12 If the body is viewed as continuous, the sum in equation (7.32) can be replaced by an integral over the volume of the body, with m_α replaced by $\rho(x)dx$, where $\rho(x)$ is the density at x, but as this makes no essential difference to what follows we consider the simpler finite sum.

If we take the dependent variables in $L\,dt$ to be the components of the $\mathbf{x}_\alpha$, then the Euler–Lagrange equations give simply that $\ddot{\mathbf{x}}_\alpha = 0$. But these are not the equations for a rigid body! For a rigid body, and this defines 'rigid body', we have

$$\mathbf{x}_\alpha(t) = g(t)\mathbf{x}_\alpha(0) + \mathbf{a}(t) \tag{7.34}$$

where $g(t)$ is a path in $SO(3)$ and $\mathbf{a}(t)$ a path in $\mathbb{R}^3$, and both $g(t)$ and $\mathbf{a}(t)$ are the same for all particles. *The Euler–Lagrange equations for a rigid body are those obtained by considering the parameters of g and $\mathbf{a}$ to be the dependent variables in the Lagrangian.*

We now note that the inner product $\langle \mathbf{x}, \mathbf{x} \rangle$ can be written, considering $\mathbf{x}$ to be a column vector, as

$$\langle \mathbf{x}, \mathbf{x} \rangle = \mathbf{x}^T \mathbf{x} = \text{trace}\left(\mathbf{x}\mathbf{x}^T \right).$$

The right hand form is the most convenient for the mathematics. Using it, the Lagrangian is written as

$$L\,dt = \sum \tfrac{1}{2} m_\alpha \text{trace}\left(\mathbf{x}_\alpha \mathbf{x}_\alpha^T \right) dt. \tag{7.35}$$

Setting equation (7.34) into (7.35), we obtain

$$L = \text{trace}\left(\dot{g}(t) \left[\sum_\alpha m_\alpha \mathbf{x}_\alpha(0)\mathbf{x}_\alpha(0)^T \right] \dot{g}(t)^T \right) + \left\langle \dot{\mathbf{a}}(t), \dot{\mathbf{a}}(t) \right\rangle$$

where we have used equation (7.33) to eliminate two terms. Denote by

$$\mathcal{M} = \sum_\alpha m_\alpha \mathbf{x}_\alpha(0)\mathbf{x}_\alpha(0)^T$$

the 3×3 matrix appearing in the Lagrangian; we have both $\mathcal{M}^T = \mathcal{M}$ and $\mathrm{d}\mathcal{M}/\mathrm{d}t = 0$. This is not quite the classical 'inertia tensor' but it is the matrix that appears naturally so we stick with it.

The only time dependent variables now appearing in the Lagrangian are the group parameters appearing in g and $\mathbf{a}$, and these are the dependent variables of the dynamical problem.

The Euler–Lagrange system for the the vector $\mathbf{a}(t)$ is trivially seen to be $\ddot{\mathbf{a}} = 0$, so $\mathbf{a}(t) = t\mathbf{a}_1 + \mathbf{a}_0$ where $\mathbf{a}_i$ are constant vectors. The more interesting task is to obtain the Euler–Lagrange system for $g(t)$. We first note that $\mathrm{trace}(A) = \mathrm{trace}(B^{-1}AB)$ and hence setting $\mathcal{A} = g^{-1}\dot{g}$ and noting that for $g \in SO(3)$ we have $g^T = g^{-1}$ and hence $\mathcal{A}^T = -\mathcal{A}$, we obtain

$$\mathcal{L}[g] = -\int \mathrm{trace}\,(\mathcal{A}\mathcal{M}\mathcal{A})\,\mathrm{d}t. \tag{7.36}$$

We now have a Lagrangian that is invariant for the action of $G = SO(3)$ on itself given by left multiplication, since for the constant element $h \in G$, we have $g^{-1}\dot{g} = (hg)^{-1}\mathrm{d}(hg)/\mathrm{d}t$. The components of $\mathcal{A}$ are the differential invariants of the group action. We now seek the Euler–Lagrange system directly in terms of these invariants.

Let ϵ be the dummy variable used to calculate the variation. Setting $\mathcal{B} = g^{-1}g_\epsilon$, we have that

$$\frac{\mathrm{d}}{\mathrm{d}\epsilon}\mathcal{A} = \frac{\mathrm{d}}{\mathrm{d}t}\mathcal{B} + [\mathcal{A}, \mathcal{B}]$$

where the last summand is the standard matrix Lie bracket. This gives us the syzygy we need to employ the methods developed above. We then have,

$$\frac{\mathrm{d}}{\mathrm{d}\epsilon}\mathcal{L} = -\int \mathrm{trace}\left(\frac{\mathrm{d}}{\mathrm{d}\epsilon}\mathcal{A}\mathcal{M}\mathcal{A} + \mathcal{A}\mathcal{M}\frac{\mathrm{d}}{\mathrm{d}\epsilon}\mathcal{A}\right)$$

$$= -\int \mathrm{trace}\left(\left(\frac{\mathrm{d}}{\mathrm{d}t}\mathcal{B} + [\mathcal{A}, \mathcal{B}]\right)\mathcal{M}\mathcal{A} + \mathcal{A}\mathcal{M}\left(\frac{\mathrm{d}}{\mathrm{d}t}\mathcal{B} + [\mathcal{A}, \mathcal{B}]\right)\right)$$

$$= -\int \mathrm{trace}\left(\mathcal{B}\left[-\mathcal{M}\frac{\mathrm{d}}{\mathrm{d}t}\mathcal{A} + \mathcal{M}\mathcal{A}^2 - \frac{\mathrm{d}}{\mathrm{d}t}\mathcal{A}\mathcal{M} - \mathcal{A}^2\mathcal{M}\right]\right)\mathrm{d}t$$
$$+ \mathrm{B.T.s}$$

where we have used $\mathrm{trace}(AB) = \mathrm{trace}(BA)$ for any square matrices A and B and

$$\mathrm{trace}\left(\frac{\mathrm{d}A}{\mathrm{d}t}\right) = \frac{\mathrm{d}\,\mathrm{trace}(A)}{\mathrm{d}t}.$$

A simple calculation shows the following.

Lemma 7.3.13 *If $\mathcal{B}$ and X are 3×3 matrices, and $\mathrm{trace}(\mathcal{B}X) = 0$ for all skew symmetric matrices $\mathcal{B}$, then X is symmetric, $X^T = X$.*

Applying the Lemma and noting $\mathcal{M}^T = \mathcal{M}$ and $\mathcal{A}^T = -\mathcal{A}$, we conclude that

$$-\mathcal{M}\frac{\mathrm{d}}{\mathrm{d}t}\mathcal{A} + \mathcal{M}\mathcal{A}^2 - \frac{\mathrm{d}}{\mathrm{d}t}\mathcal{A}\mathcal{M} - \mathcal{A}^2\mathcal{M} = 0$$

is the desired Euler–Lagrange system for $\mathcal{A}$. We can simplify the equation by noting that since $\mathcal{M}$ is symmetric, we can write it in the form

$$\mathcal{M} = U^T \Lambda U, \qquad \Lambda = \begin{pmatrix} \lambda_1 & 0 & 0 \\ 0 & \lambda_2 & 0 \\ 0 & 0 & \lambda_3 \end{pmatrix}, \qquad U \in SO(3)$$

where U is the matrix whose columns are the orthonormal eigenvectors of $\mathcal{M}$. Setting $\bar{A} = U^T \mathcal{A} U$ we have then

$$-\Lambda\frac{\mathrm{d}}{\mathrm{d}t}\bar{A} + \Lambda\bar{A}^2 - \frac{\mathrm{d}}{\mathrm{d}t}\bar{A}\Lambda - \bar{A}^2\Lambda = 0. \tag{7.37}$$

Finally we note that since $\bar{A}$ is skew symmetric, we have

$$\bar{A} = \begin{pmatrix} 0 & K_3 & -K_2 \\ -K_3 & 0 & K_1 \\ K_2 & -K_1 & 0 \end{pmatrix}$$

and we can write the Euler–Lagrange equations in terms of the K_i and the eigenvalues λ_i by considering the components of equation (7.37). To obtain the classical free rigid body equations, however, we need to convert from $\mathcal{M}$ to the classical inertia tensor $\mathcal{J}$; the relationship of $\mathcal{J}$ to $\mathcal{M}$ is

$$\mathcal{J} = \chi \begin{pmatrix} 1 & 0 & 0 \\ 0 & 1 & 0 \\ 0 & 0 & 1 \end{pmatrix} - \mathcal{M}, \qquad \chi = \sum_\alpha m_\alpha \langle \mathbf{x}_\alpha(0), \mathbf{x}_\alpha(0) \rangle.$$

Thus, $\mathcal{J}$ and $\mathcal{M}$ are both diagonalised by the same matrix U and the eigenvalues of $\mathcal{J}$, usually denoted as I_1, I_2 and I_3, are related to the λ_i by

$$I_j = \chi - \lambda_j, \qquad \chi = \lambda_1 + \lambda_2 + \lambda_3.$$

Thus (K_1, K_2, K_3) is indeed what is known as 'the angular velocity of the body' and the equations we obtain are the classical ones,

$$\frac{\mathrm{d}K_1}{\mathrm{d}t} = \frac{I_3 - I_2}{I_1} K_2 K_3$$

$$\frac{\mathrm{d}K_2}{\mathrm{d}t} = \frac{I_1 - I_3}{I_2} K_1 K_3$$

$$\frac{\mathrm{d}K_3}{\mathrm{d}t} = \frac{I_2 - I_1}{I_3} K_1 K_2.$$

Once $\mathcal{A}$ has been obtained, the rotational motion of the body, given by $g(t)$, is obtained by integrating $g^{-1}\dot{g} = \mathcal{A}$, that is,

$$\dot{g} = \mathcal{A}g$$

for g. Integration methods that guarantee the numerical solution g is an element of $SO(3)$ are available, see Celledoni *et al.* (2008).

It is clear that aspects of the method above apply to Lagrangians that are functions of $g^{-1}g_t$ for any Lie group; we leave the investigation of these as an open problem.

7.4 Moving frames and Noether's Theorem, the main course

In Section 7.2.1, we saw that the first integrals of a particular Lagrangian invariant under a Euclidean group, as obtained using Noether's Theorem, could be written using the inverse of a matrix representation of a frame, and a vector of invariants. In this section, we give a general formula for the vector of invariants and prove that the particular matrix representation is always the Adjoint representation, discussed in Section 3.3.

The theorem is important because it reveals the *structure* of the first integrals, in terms of how the surfaces that correspond to them are arranged into equivalence classes under the group action.

Theorem 7.4.1 *Suppose we have a frame ρ for the action $G \times M \to M$, where $M = J^N((x, u^\alpha))$ and x is invariant, such that the generating invariants are κ^α, one for each dependent variable.*[†] *Let the invariant Lagrangian $\int L(\kappa^\alpha, \kappa_x^\alpha, \ldots) \, dx$ be given. Introduce the dummy variable t to effect the variation, and suppose that*

$$\frac{d}{dt} L \, dx = \sum E^\alpha(L) I_2^\alpha \, dx + \frac{d}{dx} \left[\sum_{\alpha, J} I_{2J}^\alpha C_J^\alpha \right], \tag{7.38}$$

for some expressions C_J^α in the invariants, is obtained after integration by parts; here $I_2^\alpha = \iota(u_t^\alpha)$, J is an index with respect to the x variable only, $I_{2J}^\alpha = \iota(u_{tJ}^\alpha)$. Note that equation (7.38) defines the vector $C^\alpha = (C_J^\alpha)$. Let $(a_1, a_2, \ldots, a_r)$ be coordinates of G about the identity e and let $\mathbf{v}_i$, $i = 1, \ldots, r$ be the corresponding infinitesimal vector fields. Let Ad be the Adjoint representation of G with respect to these vector fields. Denote the matrices of infinitesimals, one

---

[†] This requirement will be relaxed later.

for each dependent variable, as

$$\Omega^\alpha = a_j \left(\cdots \quad \left. \frac{\partial}{\partial a_j} \right|_e g \cdot u_J^\alpha \begin{array}{c} u_J^\alpha \\ \vdots \\ \\ \vdots \end{array} \cdots \right)$$

and set $\Omega^\alpha(I) = \rho \cdot \Omega^\alpha$ *to be the componentwise invariantisation of* Ω^α. *Then the r first integrals obtained via Noether's Theorem can be written in vector form as*

$$Ad(\rho)^{-1} \sum_\alpha \Omega^\alpha C^\alpha = \mathbf{c}.$$

Proof The key idea is to conflate the dummy variable t with a group parameter a_i, say. The first implication is that since the Lagrangian is invariant, the right hand side of equation (7.38) is identically zero; this follows from differentiating both sides of $g \cdot L \, \mathrm{d}x = L \, \mathrm{d}x$ with respect to a_i. Thus on solutions of the Euler–Lagrange equations $E^\alpha(L) = 0$, we have that

$$c_i = \sum_{\alpha, J} I_{2J}^\alpha C_J^\alpha \tag{7.39}$$

where c_i is a constant. The remainder of the proof consists in expressing the I_{2J}^α in terms of the infinitesimals relevant to the group parameter a_i.

We first consider the case of a single dependent variable, and show how to generalise to more than one u^α at the end of the proof.

First note that applying $Ad(g)$ to

$$\frac{\mathrm{d}}{\mathrm{d}t} = \sum_K u_{tK} \frac{\partial}{\partial u_K}$$

at $g = \rho$ yields

$$(I_2 \, I_{12} \cdots) = (u_t \, u_{xt} \, \cdots) \left(\frac{\partial(\widetilde{u}, \widetilde{u}_x, \ldots)}{\partial(u, u_x, \ldots)} \right)^T \bigg|_{\text{frame}}. \tag{7.40}$$

Second, since t is identified with a group parameter a_i, the derivative term u_{tK} is identified with the infinitesimal $\phi_{i,K}$ (taking $t = 0$), so that $(u_t \, u_{xt} \, \cdots)$ is the ith row in the matrix of (uninvariantised) infinitesimals for u, which is $\Omega(z)$, where $z = (u, u_x, \ldots)$. Thus, equation (7.40) can be written as

$$(I_2 \, I_{12} \cdots) = \Omega_i(z) \left(\frac{\partial \widetilde{z}}{\partial z} \right)^T \bigg|_{\text{frame}}, \tag{7.41}$$

where Ω_i is the ith row of Ω. Next we use equation (3.46), which is the matrix form of what is essentially the definition of the Adjoint representation Ad, and which we repeat here for convenience,

$$Ad(g)\Omega(z) = \Omega(\widetilde{z})\left(\frac{\partial\widetilde{z}}{\partial z}\right)^{-T}. \tag{7.42}$$

Rearranging equation (7.42) and setting g to be the frame ρ yields

$$\Omega(z)\left(\frac{\partial\widetilde{z}}{\partial z}\right)^{T}\Bigg|_{\rho} = Ad(\rho)^{-1}\Omega(I).$$

Thus for $t = a_i$ we have

$$(I_2\ I_{12}\cdots) = \left(Ad(\rho)^{-1}\Omega(I)\right)_i.$$

Inserting this last into equation (7.39) and writing $\mathbf{c} = (c_1\ c_2\ \cdots\ c_r)^T$ yields

$$\mathbf{c} = Ad(\rho)^{-1}\Omega(I)\mathcal{C}.$$

If there is more than one dependent variable, so that $u = (u^1, u^2, \ldots, u^q)$, the proof above goes through setting Ω to be the concatenated matrix

$$\Omega = \left(\Omega^1\ \Omega^2\ \cdots\ \Omega^q\right),$$

$\mathcal{C}$ to be the concatenated vector,

$$\mathcal{C} = \begin{pmatrix} \mathcal{C}^1 \\ \mathcal{C}^2 \\ \vdots \\ \mathcal{C}^q \end{pmatrix},$$

and z to be $(z^1, z^2, \ldots, z^q)$ where $z^\alpha = (u^\alpha, u_x^\alpha, \ldots)$. $\qquad\square$

Example 7.4.2 For the action of $SL(2)$ on $(x, u(x))$ given by

$$\widetilde{x} = x, \qquad \widetilde{u} = \frac{au + b}{cu + d}, \qquad ad - bc = 1$$

with normalisation equations $\widetilde{u} = \widetilde{u_{xx}} = 0$, $\widetilde{u_x} = 1$, the generating invariant is $\iota(u_{xxx}) = \{u; x\}$, the Schwarzian derivative. For Lagrangians of the form $\int L(\sigma, \sigma_x, \ldots)\,dx$ where $\sigma = \{u; x\}$, the result is

$$\mathbf{c} = \underbrace{\begin{pmatrix} a^2 & -ac & -c^2 \\ -2ab & ad + bc & 2cd \\ -b^2 & bd & d^2 \end{pmatrix}}_{Ad(g)^{-1}}\Bigg|_{\text{frame}} \begin{pmatrix} \dfrac{\partial^2}{\partial x^2}E^\sigma(L) + \sigma E^\sigma(L) \\[6pt] -2\dfrac{\partial}{\partial x}E^\sigma(L) \\[6pt] -2E^\sigma(L) \end{pmatrix} \tag{7.43}$$

where recall the frame is

$$a = \frac{1}{\sqrt{u_x}}, \qquad b = -\frac{u}{\sqrt{u_x}}, \qquad c = \frac{u_{xx}}{2(u_x)^{3/2}}, \qquad ad - bc = 1.$$

The first row yields the first integral with respect to the translation parameter b, the second yields that with respect to the scaling parameter a and the third is with respect to the projective parameter c. Thus, the precise form of the Adjoint representation in (7.43) is with respect to the basis $\mathbf{v}_b$, $\mathbf{v}_a$, $\mathbf{v}_c$ in that order and hence this is the order that needs to be taken in calculating the rows of the invariantised infinitesimal matrix. The details of the calculation for the vector of invariants, as given by Theorem 7.4.1 for the Lagrangian $L(\sigma, \sigma_x, \sigma_{xx})\,dx$, are as follows. The invariantised infinitesimal matrix needed is

$$\Omega(I) = \begin{array}{c} \\ b \\ a \\ c \end{array} \overset{\begin{array}{ccccc} u & u_x & u_{xx} & u_{xxx} & u_{xxxx} \end{array}}{\left(\begin{array}{ccccc} 1 & 0 & 0 & 0 & 0 \\ 0 & 2 & 0 & 2\sigma & 2\sigma_x \\ 0 & 0 & -2 & 0 & -8\sigma \end{array} \right)}.$$

With

$$I_{2J} = \iota(u_{t\underbrace{x \cdots x}_{|J| \text{ terms}}}),$$

the boundary terms coming from the integration by parts process after performing the variational derivative on the Lagrangian is,

$$\text{B.T.s} = \sum_J I_{2J} \mathcal{C}_J$$

where

$$\mathcal{C} = \begin{array}{c} I_2 \\ I_{12} \\ I_{112} \\ I_{1112} \\ I_{11112} \end{array} \left(\begin{array}{c} \dfrac{d^2}{dx^2} E^\sigma(L) + \sigma E^\sigma(L) \\[2ex] -\dfrac{d}{dx} E^\sigma(L) - \left(\dfrac{\partial L}{\partial \sigma_x} - \dfrac{d}{dx}\dfrac{\partial L}{\partial \sigma_{xx}} \right)\sigma - \dfrac{\partial L}{\partial \sigma_{xx}}\sigma_x \\[2ex] E^\sigma(L) - 4\dfrac{\partial L}{\partial \sigma_{xx}}\sigma \\[2ex] \dfrac{\partial L}{\partial \sigma_x} - \dfrac{d}{dx}\dfrac{\partial L}{\partial \sigma_{xx}} \\[2ex] \dfrac{\partial L}{\partial \sigma_{xx}} \end{array} \right),$$

and all other $\mathcal{C}_J$ being zero. It can be seen that $\Omega(I)\mathcal{C}$ yields the vector of invariants above.

The next exercise is the analogue of Exercise 7.2.11 for the above Example.

Exercise 7.4.3 Consider equation (7.43) as three equations for u, u_x and u_{xx} by considering σ to be known once the Euler–Lagrange equation is solved. Show that there is an equation relating the c_i to a first integral of the Euler–Lagrange equation. Indeed, if the vector of invariants is written as $(I_1, I_2, I_3)^T$, show that

$$4I_1 I_3 + I_2^2 = 4c_1 c_3 + c_2^2.$$

Show that to obtain u, it is necessary to integrate only a first order Ricatti equation where the only non-constant coefficient can be incorporated into the independent variable, specifically,

$$I_3 u_x = -c_1 u^2 + c_2 u + c_3.$$

Exercise 7.4.4 Consider the one parameter group action,

$$\alpha \cdot \begin{pmatrix} x \\ y \end{pmatrix} = \begin{pmatrix} \cosh\alpha & \sinh\alpha \\ \sinh\alpha & \cosh\alpha \end{pmatrix} \begin{pmatrix} x \\ y \end{pmatrix}$$

in the plane. Show that the infinitesimal vector field is $\mathbf{v} = y\partial_x + x\partial_y$ and that the adjoint action $\mathcal{A}d(\alpha)(\mathbf{v}) = \mathbf{v}$ for all α. Conclude that the first integral of the Euler–Lagrange equations of any Lagrangian for curves $(x(t), y(t))$ in the plane,

$$\int L(t, x, y, \dot{x}, \dot{y}, \dots)\, dt$$

invariant under this action is itself invariant. Verify the result on an example.

Further results appear in Gonçalves and Mansfield (2009). It is not hard to see how to generalise Theorem 7.4.1. If there is more than one generating invariant per dependent variable, then any syzygies between them need to be incorporated into the Lagrangian as constraints, each with its own Lagrange multiplier. Similarly, if there is an arc length constraint, or a companion equation is used following reparametrisation, then these need to be incorporated in the same way. The Lagrange multipliers will appear in $\mathcal{C}$ and need to be eliminated using (first integrals of) the Euler–Lagrange equations.

Finally, we note that we do not give a formula for the boundary terms appearing in equation (7.38) in terms of the generating invariants, leaving this part to be achieved by symbolic computation. Converting the methods given here to code that can take as input Lagrangians that are polynomial (say) in their arguments is straightforward, once the symbolic differentiation formulae have been coded.

References

Ablowitz, M. J. and Clarkson, P. A. (1991). *Solitons, Nonlinear Evolution Equations and Inverse Scattering*, London Mathematical Society Lecture Note Series **149**. Cambridge: Cambridge University Press.

Akivis, M. A. and Rosenfeld, B. A. (1993). *Élie Cartan (1869–1951)*, Translations Mathematical Monographs, **123**. Providence, RI: American Mathematical Society.

Bluman, G. W. and Cole, J. D. (1974). *Similarity Methods for Differential Equations*, Applied Mathematical Sciences, **13**. New York: Springer.

Bluman, G. W. and Kumei, S. (1989). *Symmetries and Differential Equations*, Applied Mathematical Sciences, **81**. New York: Springer.

Boutin, M. (2002). On orbit dimensions under a simultaneous Lie group action on n copies of a manifold, *J. Lie Theory*, **12**, 191–203.

Boutin, M. (2003). Polygon recognition and symmetry detection, *Found. Comput. Math.*, **3**, 227–271.

Brown, R. E. (1993). *A Topological Introduction to Nonlinear Analysis*. Boston, MA: Birkhäuser.

van Brunt, B. (2004). *The Calculus of Variations*. New York: Springer.

Cantwell, B. J. (2002). *Introduction to Similarity Analysis*, Cambridge Texts in Applied Mathematics. Cambridge: Cambridge University Press.

Carminati, J. and Vu, K. (2000). Symbolic computation and differential equations: Lie symmetries, *J. Symbolic Comput.*, **29**, 95–116.

Cartan, E. (1953). *Oeuvres Complètes*. Paris: Gauthier–Villars.

Cerda, E., Mahadevan, L. and Pasini, J. M. (2004). The elements of draping, *Proc. Natl. Acad. Sci.*, **101**, 1806–1810.

Celledoni, E., Fasso, F., Saefsroem, S. and Zanna, A. (2008). The exact computation of the free rigid body motion and its use in splitting methods, *SIAM J. Sci. Comput.*, **30** (4), 2084–2112.

Chan, T., Kang, S. H. and Shen, J. (2002). Euler's elastica and curvature based inpaintings, *SIAM J. Appl. Math.*, **63** (2), 564–592.

Cheh, J., Olver, P. J. and Pohjanpelto, J. (2005). Maurer–Cartan equations for Lie symmetry pseudo-groups of differential equations, *J. Math. Phys.*, **46** (2) 023504.

Choquet-Bruhat, Y. and DeWitt-Morette, C., with Dillard-Bleick, M. (1982). *Analysis, Manifolds and Physics*. Amsterdam: North-Holland.

Clarkson, P. A. and Mansfield, E. L. (1993). Symmetry reductions and exact solutions of a class of nonlinear heat equations, *Physica D*, **70**, 230–288.

Clarkson, P. A. and Olver, P. J. (1996). Symmetry and the Chazy equation, *J. Differential Equations*, **124** (1), 225–246.

Courant, R. and Hilbert, D. (1953). *Methods of Mathematical Physics*. New York: Wiley.

Fässler, A. and Stiefel, E. (1992). *Group Theoretical Methods and their Applications*, translated by B. D. Wang. Boston, MA: Birkhäuser.

Fels, M. and Olver, P. J. (1998). Moving coframes I, *Acta Appl. Math.*, **51**, 161–213.

Fels, M. and Olver, P. J. (1999). Moving coframes II, *Acta Appl. Math.*, **55**, 127–208.

de Graaf, W. A. (2000). *Lie Algebras: Theory and Algorithms*, North-Holland Mathematical Library, **56**. Amsterdam: Elsevier.

Gilmore, R. (1974). *Lie Groups, Lie Algebras and Some of their Applications*, New York: Wiley. Reprinted by Dover Publications, Mineola, New York, 2002.

Gonçalves, T. M. N. and Mansfield, E. L. (2009). Moving frames and Noether's theorem.

Hirsch, M. W. (1976). *Differential Topology*. New York: Springer.

Hoschek, J. and Lasser, D. (1993). *Fundamentals of Computer-Aided Geometric Design*, translated by L. L. Schumaker. Natick, MA: AK Peters.

Hubert, E. (2000). Factorization free decomposition algorithms in differential algebra, *J. Symbolic Comput.*, **29** (4–5), 641–662.

Hubert, E. (2005). Differential algebra for derivations with nontrivial commutation rules, *J. Pure Appl. Algebra*, **200** (1–2), 163–190.

Hubert, E. (2009a). Differential invariants of a Lie group action: syzygies on a generating set, *J. Symbolic Comput.*, **44** (4), 382–416.

Hubert, E. (2009b). Generation properties of Maurer–Cartan invariants. Preprint [hal:inria-00194528]

Hubert, E. and Kogan, I. A. (2007a). Smooth and algebraic invariants of a group action. Local and global constructions, *Foundations Comput. Math.*, **7** (4), 345–383.

Hubert, E. and Kogan I. A. (2007b). Rational invariants of a group action. Construction and rewriting, *J. Symbolic Comput.*, **42** (1–2), 203–217.

Hydon, P. H. (2000). *Symmetry Methods for Differential Equations: a Beginner Guide*, Cambridge Texts in Applied Mathematics. Cambridge: Cambridge University Press.

Ibragimov, N. H. (1992). Group analysis of ordinary differential equations and the invariance principle in mathematical physics, *Russian Math. Surveys*, **47**, 89–156.

Ibragimov, N. H. (ed.) (1996). *CRC Handbook of Lie Group Analysis of Differential Equations*, Volume 3, *New Trends in Theoretical Developments and Computational Methods*. Boca Raton, FL: CRC Press.

Iserles, A., Munthe-Kaas, H., Nørsett, S. P. and Zanna, A. (2000). Lie-group methods, *Acta Numerica*, **9**, 215–365.

Kogan, I. A. (2000a). Inductive construction of moving frames. In *The Geometrical Study of Differential Equations* (Washington, DC, 2000), Contemporary Mathematics, **285**, pp. 157–170. Providence, RI: American Mathematical Society.

Kogan, I. A. (2000b). Inductive Approach to Cartan's Moving Frame Method with Applications to Classical Invariant Theory. *PhD Thesis*, University of Minnesota.

Kogan, I. A. and Olver, P. J. (2003). Invariant Euler–Lagrange equations and the invariant variational bicomplex, *Acta Appl. Math.*, **76**, 137–193.

MacCluer, C. R. (2005). *Calculus of Variations: Mechanics, Control and Other Applications*. Englewood Cliffs, NJ: Pearson Prentice Hall.

Mansfield, E. L. (2001). Algorithms for symmetric differential systems, *Foundations Comput. Math.*, **1**, 335–383.

Mansfield, E. L. and van der Kamp, P. (2006). Evolution of curvature invariants and lifting integrability, *J. Geometry Phys.*, **56**, 1294–1325.

Marí Beffa, G. (2004). Poisson brackets associated to invariant evolutions of Riemannian curves, *Pacific J. Math.*, **215** (2), 357–380.

Marí Beffa, G. (2007). On completely integrable geometric evolutions of curves of Lagrangian planes, *Proc. R. Soc. Edinburgh A, Math.*, **137**, 111–131.

Marí Beffa, G. (2008a). Hamiltonian evolutions of curves in classical affine geometries, *Physica D*, **238**, 100–115.

Marí Beffa G. (2008b). Projective-type differential invariants and geometric curve evolutions of KdV-type in flat homogeneous manifolds, *Ann. l'Institut Fourier*, **58** (4), 1295–1335.

Mumford, D. (1994). Elastica and computer vision. In *Algebraic Geometry and Its Applications*, ed. C. L. Bajaj, pp. 491–506. New York: Springer.

Munkres, J. R. (1975). *Topology: a First Course*. Englewood Cliffs, NJ: Prentice-Hall.

Noether, E. (1918). Invariante Variationsprobleme. *Nachr. Ges. Wiss. Goettingen, Math.-Phys. Kl.*, 235–257. An english translation is available at arXiv:physics/0503066v1 [physics.hist-ph]

Olver, P. J. (1993). *Applications of Lie Groups to Differential Equations*, second edition. New York: Springer.

Olver, P. J. (1995). *Equivalence, Invariants, and Symmetry*. Cambridge: Cambridge University Press.

Olver, P. J. (2000). Moving frames and singularities of prolonged group actions, *Selecta Math.*, **6**, 41–71.

Olver, P. J. (2001a). Joint invariant signatures, *Found. Comput. Math.*, **1**, 3–67.

Olver, P. J. (2001b). Geometric foundations of numerical algorithms and symmetry, *Applicable Algebra Eng. Commun. Comput.*, **11**, 417–436.

Olver, P. J. and Pohjanpelto, J. (2008). Moving frames for Lie pseudo-groups, *Can. J. Math.*, **60**(6), 1336–1386.

Ovsiannikov, L. V. (1982). *Group Analysis of Differential Equations*, translated by W. F. Ames. New York: Academic Press.

Saunders, D. J. (1989). *The Geometry of Jet Bundles*. Cambridge: Cambridge University Press.

Stephani, H. (1989). *Differential Equations, their Solution using Symmetries*, ed. M. MacCallum. Cambridge: Cambridge University Press.

Shemyakova, E. and Mansfield, E. L. (2008). Moving frames for Laplace invariants. In *Proceedings of International Symposium of Symbolic and Algebraic Computation, Linz, Austria, 2008*, ed. D. Jeffrey, pp. 295–302. New York: Association of Computing Machinery.

Tapp, K. (2005). *Matrix Groups for Undergraduates*, Student Mathematical Library, **29**. Providence, RI: American Mathematical Society.

Tresse, A. (1894). Sur les invariants différentiels des groupes continus de transformations, *Acta Math.*, **18**.

Whittaker, E. T. and Watson, G. N. (1952). *A Course in Modern Analysis*, fourth edition. Cambridge: Cambridge University Press.

Index

Printed in the United States
by Baker & Taylor Publisher Services